11/29/89

Native and
Cultivated Conifers
of Northeastern
North America

by **EDWARD A. COPE**

illustrated by
Bente Starcke King

COMSTOCK PUBLISHING ASSOCIATES, a division of

Native and Cultivated Conifers of Northeastern North America

A GUIDE

CORNELL UNIVERSITY PRESS *Ithaca and London*

Library of Congress Cataloging-in-Publication Data

COPE, EDWARD A., 1948–
    Native and cultivated conifers of northeastern North America.

    Bibliography: p.
    Includes index.
    1. Conifers—Northeastern States—Identification.
2. Conifers—Canada, Eastern—Identification.
3. Ornamental conifers—Northeastern States—
Identification.   4. Ornamental conifers—Canada,
Eastern—Identification.   I. Title.
QK494.C67   1986   585'.20974   85-24338
ISBN 0-8014-1721-X (alk. paper)
ISBN 0-8014-9360-9 (pbk.: alk. paper)

**Dedicated to Joshua A. Cope, Extension Forester,
Cornell University, 1924–1950.**

First published 1986 by Cornell University Press.
Comstock|Cornell Paperbacks edition first published 1986.

Printed in the United States of America

*The paper in this book is acid-free and meets the guidelines for
permanence and durability of the Committee on Production Guidelines
for Book Longevity of the Council on Library Resources.*

Contents

Acknowledgments

I thank the following people, whose assistance made this book possible: James Cope; Peter Hyypio; David Young, director of the Bailey Hortorium; Robert Fincham, president of the American Conifer Society; Jean Reed; Sherry Vance; and Bente Starcke King.

All drawings were made by Bente Starcke King. James Cope contributed photographs of *Juniperus virginiana, Larix laricina, Microbiota decussata, Picea omorika, Pinus strobus,* and *Taxodium distichum* from his arboretum in Centerville, Indiana. The other photographs were taken by the author at Westtown School Arboretum, Scott Foundation Arboretum at Swarthmore College, Tyler Arboretum, Cornell University campus, and Secrest Arboretum except for that of *Sequoia sempervirens,* which was taken in Humboldt County, California.

EDWARD A. COPE

Ithaca, New York

**Native and
Cultivated Conifers
of Northeastern
North America**

Introduction

Conifers, the major group of evergreen plants in northeastern North America, dominate the winter landscape, and in the growing season their shapes and textures contrast with those of other woody plants. Persistent green, needlelike leaves usually distinguish conifers from flowering plants, although some flowering plants have evergreen needlelike leaves and a few conifers are deciduous. In addition, some conifers produce broad leaves, but these species are not found in the temperate climate of the Northeast. The reproductive structure—the cone with exposed seeds—is the most important feature that differentiates the conifers from the flowering plants.

Present-day conifers date back to Paleozoic times, some 300 million years ago. As a group they dominated the Earth's flora during much of the dinosaur age (Mesozoic era) until flowering plants rose to dominance in the late Mesozoic, 125 million years ago, a position that the flowering plants have occupied ever since. The conifers, however, have survived and in some parts of the world, such as the mountain and coastal areas of western North America and the boreal forests of northern North America, are the most prominent of the two types of plant.

In northeastern North America, conifers account for only 18 of the native plant species, yet populations of these species comprise a disproportionately large part of the present native flora. Extensive forests of white pine, hemlock, and balsam fir preceded European colonization and the timber industry that followed. The same species, in rejuvenated forests, remain the most evident native conifers in the Northeast today and have been extensively cultivated, together with many other conifers, because of their many cultural uses, the most important of which are lumber, Christmas trees, and ornamentals. Several introduced species, notably Norway spruce and Scotch pine, are well established, seeding freely and growing easily without the aid of man.

Purpose

The present book seeks to provide a means of identifying the native and cultivated conifers of northeastern North America, using vegetative features (leaves, twigs, and buds), simple terms, and many illustrations. Some cultivated species grown in arboreta or in botanical gardens are unlikely to be found in a yard, park, or nursery. A few of these have not been included. Information on such rare exotics can be found in the Selected References. Twenty-seven genera and 130 species have been included here.

Most manuals rely heavily on cone characteristics for the identification of conifers. This, of course, is the surest method, because conifer taxonomy is based largely upon differences in reproductive structure. Frequently, however, only branches and leaves are available, inasmuch as cones are produced late in the life of most conifers and then often appear only near the top of a tall tree. The keys and other aids that I present here do not depend on reproductive material for identification of the plant in question. Unlike many available publications that encompass only native trees, this work includes cultivated species that are usually omitted. It should also serve as an extensive reference list of named cultivars.

Geographical Area

The territory covered is roughly eastern Canada and the northeastern fourth of the United States, encompassing the region from Maine south to the southern border of Pennsylvania, west to Kansas, and north to North Dakota. Because so many cultivated species have been included, the manual may be useful in an area broader than that just defined (and particularly in western North America and other temperate regions of the world).

Method

The book provides a key to genera with many illustrations; a list of the genera in alphabetical order with brief descriptions, an illustration, and a photograph; and a key to the species (when there is more than one) of each genus. The species are then listed alphabetically with a short description, followed by the cultivars (if any) with their descriptions. In species with large numbers of cultivars, groups of names by selected characters follow the list of described cultivars. Appendix 1 provides character groups for cultivars of all the species that have been included in this book. Appendix 2 presents a table showing conifer families and genera of the world, with their geographic distribution. Appendix 3 illustrates representative cones and seeds of all genera discussed. Appendix 4 shows representative twigs of the same 27 genera. Finally, I have listed some valuable and informative although mostly dated references dealing with conifers.

My discussions of species are brief, since detailed descriptions and photographs may be found in a variety of other publications (see Selected References). Common names are presented according to established procedure, with the preferred name listed first (Little, 1979). An asterisk in the descriptions and keys indicates species commonly cultivated in northeastern North America. A plus sign indicates species that are native to the region. All other species may well appear in a botanical garden, arboretum, or other public place but are probably not to be found in many nurseries or in a home-owner's yard.

Nearly all the keys use vegetative features (leaves, twigs, and buds) that can be seen with the naked eye; in a very few cases a hand lens of low power is required.

Drawings have been inserted to facilitate use of the keys. Descriptions of reproductive material or deciduous habit have occasionally been incorporated when they facilitate quick determination, but this information is not essential for the eventual identification of the species covered in the keys. (The exception is an intergeneric cross, ×*Cupressocyparis,* in which case both vegetative and reproductive characters are required for identification.) When the keys are consulted, at least second-year or older leaves and branches should be used (that is, not the new growth or young shoots) unless the key step indicates otherwise. The characteristics of young leaves and twigs in the first year of growth often differ from those of older parts of the plant. Occasionally a cultivar deviates so radically in growth habit or other features from typical characters of the species that the species key will not help identify it.

I examined both living material and herbarium specimens. Some of the references that I have listed were useful in the preparation of the keys and descriptions. In nearly all cases the illustrations were made using living material.

Cultivars

Many horticultural varieties of conifers have been developed and can be seen in home plantings, arboreta, nurseries, and parks. Most of the categories of plants below species rank in this treatment are cultivars—which are designated by single quotation marks and have been defined as an assemblage of "cultivated plants which is clearly distinquished by any characters (morphological, physiological, cytological, chemical or others), and which, when reproduced (sexually or asexually), retains its distinguishing characters" (Brickell, 1980, p. 12). The designations *var.* (variety), *ssp.* or *subsp.* (subspecies), and *f.* (forma) have more precise botanical meaning and further subdivide species.

I have included a total of 2669 cultivars in this book. The lists with each species are not complete, although I attempted to make them as exhaustive as possible, at least with respect to cultivars found in northeastern North America. Some of the conifers listed, such as European cultivars or plants from this country that are known by older names, may no longer exist or may be unavailable in this region.

I have tried to keep descriptions brief partly to include all the cultivar names in a manageable form. The cultivar descriptions usually note only features that deviate from the normal species characteristics. In some cases the descriptions are short because accurate information is not available or because information is conflicting.

Where no information is available, a cultivar name simply appears without description.

I have minimized the number of terms for each character describing a cultivar in the attempt to cover all possibilities for variation. Each nurseryman, consumer, and author has differing notions of the various adjectives used to describe plants. In this work I have usually replaced terms such as *silver, silvery blue, bluish, bluish white, whitish, yellowish, gold, golden yellow,* and *pale yellow* with *blue, white,* and *yellow.* Since growth rates in various types of soils and climatic conditions vary greatly, I have employed only two general terms for departures from the normal rate of growth of the species. The terms *dwarf* and *slow* are intended to describe plants growing so slowly that in 15 years they would not be expected to achieve a height greater than one meter (three feet) and two meters (six feet), respectively. Any actual figures for growth rate that appear in this book are to be used only as guidelines and reflect the experience of other observers. The terms *columnar, conical, rounded,* and *spreading,* refer to growth habit. Some shapes encompassed by each of these terms are illustrated on page 17.

Many cultivar names are regarded by conifer specialists and nurserymen as synonyms, or designations that have become established despite the existence of earlier names for the same plants. In this work synonyms are listed with the proper name, which appears after the phrase *same as.* In the case of plants that have different names but are very nearly identical, the words *similar to* precede the cultivar name under which the description appears.

In some species, cultivars have been grouped under types of growth habit, growth rate, or color form, a practice developed by Gerd Krüssmann in his publications on conifers. I hope that this approach can be further advanced in future manuals and can help eliminate some of the confusing nomenclature that pertains to conifer cultivars. To be useful, a named and described cultivar should retain its distinguishing characteristics for at least 15 years. It does not always do so because most conifer growers do not wait such a lengthy time to see results of their propagation trials before marketing a new cultivar. The most careful growers allow enough years of trial to permit a reasonable projection of growth characteristics 15 years ahead, but even they can be fooled. Furthermore, some plants, usually rooted or grafted cuttings, develop different characteristics after several or even many years so that plants with the identical cultivar name may look very different.

Information included in the cultivar lists in this work was derived from nursery catalogues and reflects my observations and sometimes those of private collectors. It has been assembled for quick reference and to provide a basis for further progress toward more complete and accurate descriptions. The list is American in its emphasis, and wherever information about a cultivar name is conflicting, the American concept is used. Difficulties arise when the conflicting descriptions both come from American sources. In some cases an arbitrary decision has been made, pending further insight into the problem.

Terminology

Complex terminology has been kept to a minimum. A few terms, however, need to be defined.

alternate   arrangement whereby leaves or twigs are positioned at different longitudinal increments on an axis (stem or branch).

bract   a modified or reduced leaf between the cone scales.

facial leaves   scalelike leaves, the back surface of which is seen between the edges of the lateral leaves when viewed from above the twig. See illustration.

Facial leaf
Gland
Lateral leaf

gland   a pit, depression, groove, raised area, or bump on a leaf where resin or secretion can be produced. See illustration for one example.

glaucous   covered with a whitish substance or bloom that rubs off.

keeled   with a ridge like that along the bottom of a boat.

lateral leaves   scalelike leaves, the edges of which are seen to the right and left of the facial leaf when viewed from above the twig. See illustration.

opposite   arrangement whereby two leaves or twigs emanate from the same longitudinal position but from opposing sides of an axis (stem or branch).

spray   a branch that is at least three years old or the end of a branch that has at least three years of growth.

variegated   foliage with a pattern of two or more colors.

whorled   arrangement whereby three or more leaves or branches emanate from the same longitudinal position on an axis (stem or branch).

General terms for growth habit, or shape, are illustrated on the opposite page.

Conifer shapes

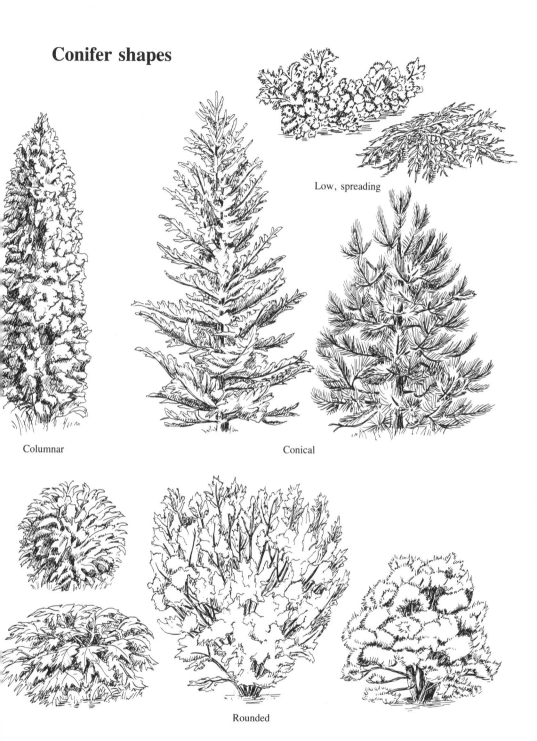

Low, spreading

Columnar

Conical

Rounded

How to Use the Keys

A key is a device that facilitates identification of an unknown object. The keys presented here are dichotomous; that is, they offer two alternatives at each step. The plant should match one of the two sets of characteristics described. The choices are numbered identically and have been set at an equal distance from the margin. Having made a choice at the first step, the reader proceeds to the next step (that is, to the next number) beneath that choice, which is further indented. This process is repeated until a dotted line leads to a name, which should be that of the plant at hand. Descriptions and drawings or photographs may be consulted to verify the identification.

It is important always to examine both alternatives at each step in the key. Several distinctive characters may be offered at each step to help the reader decide which path to follow. Drawings that illustrate the characters are positioned near the key steps. Although each is labeled with the scientific name of the plant, the reader should not assume that the plant at hand is the same as the one depicted. The drawings are intended only to illustrate particular characters described in the key steps; several species may share the character shown. When a decision seems especially difficult, one lead should be followed, then the other. Sometimes it is more efficient to work only partway through the key. Descriptions and pictures of the possibilities that remain may then make proper identification possible.

At certain difficult points in some keys, the same name can be reached through either choice. In the key to the species of *Picea,* for example, *P. asperata* can be reached under either "4. Twigs obviously hairy" or "4. Twigs lacking hairs"; twigs of *P. asperata* are normally hairy but occasionally lack hairs.

Key to Genera

1. *Abies*

2. *Larix*

3. *Metasequoia*

1. Leaves needlelike, not scalelike, at least 4 times as long as wide; twigs visible beneath the leaves (Fig. 1) (see p. 24 for other choice 1).

  2. Leaves deciduous (falling in autumn). (Use other choice if this information is not known.)

    3. Leaves in clusters on very short branches or spurs (Fig. 2) (solitary on young shoots).

      4. Leaves greater than 3.8 cm (1½ in) long and greater than 1 mm wide . . . . *Pseudolarix*
Golden-larch

      4. Leaves less than 3.8 cm (1½ in) long and 1 mm or less wide . . . . *Larix*
Larch

    3. Leaves all solitary (Fig. 3).

      5. Leaves and twigs alternate . . . . *Taxodium*
Baldcypress

      5. Leaves and twigs opposite . . . . *Metasequoia*
Dawn-redwood

  2. Leaves persistent (plants evergreen).

    6. Female reproductive structures, if present, berrylike.

      7. Berrylike structures 5–10 mm (⅓–¼ in) in diameter, blue or blue-black when ripe. . . . . *Juniperus*
Juniper

19

7. Berrylike structures at least 12 mm (½ in) in diameter, purple
   or orange-red when ripe.
   8. Berrylike structures red-orange when ripe, 13 mm (½ in)
      in diameter .... *Taxus*
                                                         Yew
   8. Berrylike structures purple when ripe, at least 15 mm (1 in)
      in diameter.
      9. Leaves with prominent midrib on upper surface, 2 broad
         (1 mm), white longitudinal bands on lower surface;
         berrylike structure stalked .... *Cephalotaxus*
                                                         Plumyew
      9. Leaves without distinct midrib, 2 narrow (0.5 mm),
         longitudinal bands on lower surface; berrylike structure not
         stalked .... *Torreya*
                                                         Torreya
6. Female reproductive structures absent or not berrylike.
   10. Leaves in clusters (Figs. 4 and 5) (solitary only on young
       shoots).
       11. Leaves in clusters or whorls of 10 or more (Fig. 5).
           12. Leaves 8–16 cm (3–6 in) long .... *Sciadopitys*
                                                         Umbrella-pine
           12. Leaves less than 7.5 cm (3 in) long.
               13. Branches from trunk arranged irregularly; leading
                   shoot drooping or hanging; spurs roughened from
                   persistent leaf sheaths (Fig. 6); mature leaves stiff,
                   sharp .... *Cedrus*
                                                         Cedar
               13. Branches in regular whorls from the trunk;
                   spurs smooth, with only rings of leaf scars remaining
                   (Fig. 7); mature leaves soft, blunt.
                   14. Leaves greater than 3.8 cm (1½ in) long and
                       greater than 1 mm wide .... *Pseudolarix*
                                                         Golden-larch
                   14. Leaves less than 3.8 cm (1½ in) long and 1 mm
                       or less wide .... *Larix*
                                                         Larch
       11. Leaves in bundles of 2, 3, 4, or 5 (Fig. 8) (solitary only
           on young shoots) .... *Pinus*
                                                         Pine
   10. Leaves all solitary (Figs. 9 and 10).
       15. Leaves greater than 8 cm (3 in) long .... *Sciadopitys*
                                                         Umbrella-pine
       15. Leaves less than 7 cm (2½ in) long.
           16. Leaves curved (Fig. 11), awl-shaped, and keeled,
               with thick ridge along the back.
               17. Leaves alternate .... *Cryptomeria*
                                                         Cryptomeria

4. *Pinus*

5. *Larix*

6. *Cedrus*

7. *Pseudolarix*

8. *Pinus*

9. *Picea*

10. *Metasequoia*

11. *Cryptomeria*

12. *Cephalotaxus*

13. *Metasequoia*

14. *Pseudotsuga*

15. *Abies*

16. *Picea*

17. *Abies*

18. *Juniperus*

19. *Cephalotaxus*

17. Leaves opposite or whorled.

    18. Leaves whorled or if opposite usually greater than 10 mm (⅜ in) long; reproductive structure berrylike .... *Juniperus*
        Juniper

    18. Leaves opposite, less than 10 mm (⅜ in) long; reproductive structure a woody cone .... *Chamaecyparis*
        False-cypress

16. Leaves straight, flat, or angled (Figs. 12 and 13), not curved, not awl-shaped, not keeled.

    19. Buds at least 2½ times as long as wide, sharp-pointed (Fig. 14), usually greater than 5 mm (¼ in) long at the end of the twigs .... *Pseudotsuga*
        Douglas-fir

    19. Buds spherical or ovoid, at most 2 times as long as wide (Fig. 15), usually less than 5 mm (¼ in) long.

        20. Leaves angular in cross section, easily rolled between fingers; twigs very rough because of persistent woody projections at the base of each leaf (Fig. 16) .... *Picea*
            Spruce

        20. Leaves flattened (Fig. 17), not easily rolled between fingers; twigs without prominent woody projections.

            21. Leaves less than 1 mm wide; twigs at the ends of the branches no more than 0.5 mm wide.

                22. Leaves alternate .... *Taxodium*
                  Baldcypress

                22. Leaves opposite .... *Metasequoia*
                  Dawn-redwood

            21. Leaves 1 mm wide or more; twigs greater than 2 mm (¹⁄₁₆ in) wide.

                23. Leaves opposite or whorled (appearing so in *Torreya*).

                  24. Leaves with a single white or gray longitudinal band on *upper* surface, edges often rolled or curled upward (Fig. 18) .... *Juniperus*
                    Juniper

                  24. Leaves with 2 white or pale longitudinal bands on *lower* surface; edges, if rolled, rolled under.

                    25. Leaves with prominent midrib on upper surface, 2 broad (1 mm), white longitudinal bands on lower surface (Fig. 19) ...... *Cephalotaxus*
                      Plumyew

                    25. Leaves without distinct midrib, with 2 narrow (0.5 mm), pale longitudinal bands on lower surface .... *Torreya*
                      Torreya

                23. Leaves alternate.

21. *Tsuga*

23. *Picea*

25. *Picea*

27. *Sequoia*

29. *Picea*

31. *Taxus*

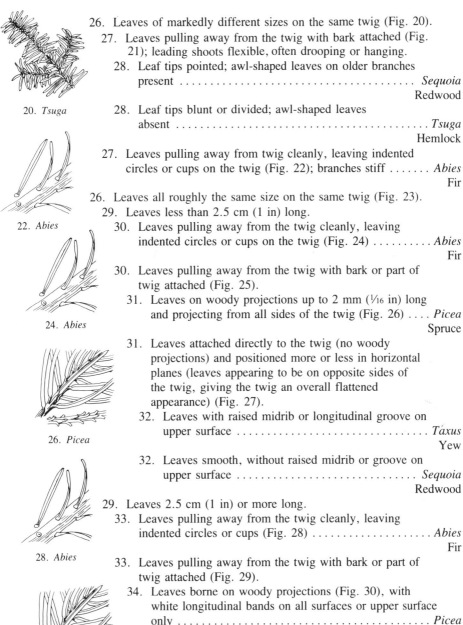

20. *Tsuga*

22. *Abies*

24. *Abies*

26. *Picea*

28. *Abies*

26. Leaves of markedly different sizes on the same twig (Fig. 20).
 27. Leaves pulling away from the twig with bark attached (Fig. 21); leading shoots flexible, often drooping or hanging.
  28. Leaf tips pointed; awl-shaped leaves on older branches present . . . . *Sequoia*
Redwood
  28. Leaf tips blunt or divided; awl-shaped leaves absent . . . . *Tsuga*
Hemlock
 27. Leaves pulling away from twig cleanly, leaving indented circles or cups on the twig (Fig. 22); branches stiff . . . . *Abies*
Fir
26. Leaves all roughly the same size on the same twig (Fig. 23).
 29. Leaves less than 2.5 cm (1 in) long.
  30. Leaves pulling away from the twig cleanly, leaving indented circles or cups on the twig (Fig. 24) . . . . *Abies*
Fir
  30. Leaves pulling away from the twig with bark or part of twig attached (Fig. 25).
   31. Leaves on woody projections up to 2 mm ($\frac{1}{16}$ in) long and projecting from all sides of the twig (Fig. 26) . . . . *Picea*
Spruce
   31. Leaves attached directly to the twig (no woody projections) and positioned more or less in horizontal planes (leaves appearing to be on opposite sides of the twig, giving the twig an overall flattened appearance) (Fig. 27).
    32. Leaves with raised midrib or longitudinal groove on upper surface . . . . *Taxus*
Yew
    32. Leaves smooth, without raised midrib or groove on upper surface . . . . *Sequoia*
Redwood
 29. Leaves 2.5 cm (1 in) or more long.
  33. Leaves pulling away from the twig cleanly, leaving indented circles or cups (Fig. 28) . . . . *Abies*
Fir
  33. Leaves pulling away from the twig with bark or part of twig attached (Fig. 29).
   34. Leaves borne on woody projections (Fig. 30), with white longitudinal bands on all surfaces or upper surface only . . . . *Picea*
Spruce
   34. Leaves directly joining the twig for some distance along the twig (Fig. 31), with white or pale longitudinal bands on lower surface only.

30. *Picea*

35. Twigs alternate or at most nearly opposite . . . . *Taxus*
Yew

35. Twigs opposite.
36. Leaves alternate, blue, wider at base, base
rounded . . . . *Cunninghamia*
China-fir

36. Leaves opposite (or at least appearing so), width at base
about equal to width at tip.
37. Leaves with prominent midrib on upper surface, with 2
broad (1 mm), white longitudinal bands on lower
surface . . . . *Cephalotaxus*
Plumyew

37. Leaves without distinct midrib on upper surface, with 2
narrow (0.5 mm), pale longitudinal bands on lower
surface . . . . *Torreya*
Torreya

32. *Chamaecyparis*

1. Leaves scalelike (Fig. 32), no greater than 2 times as long as
wide; twigs usually hidden by brown or green leaves.
38. Leaves alternate.
39. Leaves pointing forward but not pressed against twig
(Fig. 33) . . . . *Cryptomeria*
Cryptomeria

39. Leaves pressed against and covering twig
(Fig. 34) . . . . *Sequoiadendron*
Giant-sequoia

33. *Cryptomeria*

38. Leaves opposite or whorled.
40. Leaves in whorls of 4 (Fig. 35) . . . . *Calocedrus*
Incense-cedar

40. Leaves opposite (Fig. 36) or in whorls of 3.
41. Cones or berrylike structures present in material at
hand.
42. Blue, berrylike structures present . . . . *Juniperus*
Juniper

34. *Sequoiadendron*

42. Cones present.
43. Cones spherical, with the widest sides of
scales facing out or up, not overlapping.
44. Cone scales each bearing 2–5 seeds.
45. Cones more than 13 mm (½ in) in diameter,
8 scales, 5 seeds per scale . . . . . ×*Cupressocyparis*
Leyland-cypress

45. Cones 13 mm (½ in) in diameter with 6–12
scales, 2–5 seeds per scale . . . . *Chamaecyparis*
False-cypress

44. Cone scales each bearing many seeds (more
than 5) . . . . *Cupressus*
Cypress

35. *Calocedrus*

36. *Juniperus*

37. *Thujopsis*

38. *Thuja*

39. *Thuja*

40. *Thuja*

41. *Chamaecyparis*

42. *Microbiota*

43. *Chamaecyparis*

43. Cones oblong (longer than wide), widest sides of scales facing to the side and overlapping.

    46. Sprays much flattened; twigs 5–6 mm (¼ in) wide; leaves with conspicuous white markings on lower surface (Fig. 37); shrubs . . . . *Thujopsis*
            Hiba-arborvitae

    46. Sprays less flattened; twigs 2–4 mm (⅛ in) wide; leaves with or without white markings on lower surface (Fig. 38); trees or shrubs . . . . *Thuja*
            Arborvitae

41. Cones or berrylike structures absent from material at hand.

    47. Sprays cylindrical or quadrangular, in several different planes, spreading at various angles from trunk or main stems.

        48. Leaves on young shoots with a white longitudinal band on *upper* surface; common . . . . *Juniperus*
                Juniper

        48. Leaves on young shoots with white longitudinal bands on *lower* surface; rare . . . . *Cupressus*
                Cypress

    47. Sprays flattened, in several different planes or in only horizontal planes.

        49. Seam where lateral leaves meet not usually visible (Fig. 39).

            50. Visible portion of leaves greater than 3 mm (⅛ in) long and greater than 2 mm (1/16 in) wide . . . . *Thujopsis*
                Hiba-arborvitae

            50. Visible portion of leaves less than 3 mm (⅛ in) long and less than 2 mm (1/16 in) wide.

                51. Leaves blunt (Fig. 40).

                  52. Sprays mostly in horizontal planes; lateral leaves as long as wide . . . . *Thuja*
                      Arborvitae

                  52. Sprays in various planes, tending to be parallel to trunk or main stems; lateral leaves longer than wide . . . . *Platycladus*
                      Oriental-arborvitae

                51. Leaves sharp-pointed (Fig. 41).

                  53. Leaf tips on young shoots awl-shaped (Fig. 42), 1–2 mm (1/32–1/16 in) long . . . . *Microbiota*
                      Microbiota

                  53. Leaf tips on young shoots triangular, not awl-shaped (Fig. 43), usually less than 1 mm (1/32 in) long . . . . *Chamaecyparis*
                      False-cypress

44. *Chamaecyparis*

45. *Platycladus*

46. *Chamaecyparis*

49. Seam where lateral leaves meet usually visible (Fig. 44).
  54. Facial and lateral leaves with prominent groove
    (Fig. 45) . . . . *Platycladus*
                                               Oriental-arborvitae
  54. Facial and lateral leaves smooth or facial leaves only with
    narrow, inconspicuous groove or circular pit (Fig. 46).
    55. Cone scales 4–6, each bearing 2 seeds; foliage odorous
      when bruised . . . . *Chamaecyparis*
                                              False-cypress
    55. Cone scales 8, each bearing 5 seeds; foliage less odorous
      when bruised . . . . ×*Cupressocyparis*
                                              Leyland-cypress

Descriptions, Keys to Species, and Cultivar Lists

Abies grandis

Abies cilicica

Abies magnifica var. *shastensis*

Abies Mill.   Fir.

All of the approximately 50 species of firs are tall forest trees growing at high altitudes in the Northern Hemisphere. They make handsome specimens as young trees and require relatively little space for
planting, since in growth habit they are typically narrow and
spirelike. This genus is normally slower growing than the pines and
spruces. The leaves are often a rich, dark green, with two white
longitudinal bands on the lower surface. Most species have a strong
and pleasant aroma. The leaves are usually flat and blunt or divided
at the tips and leave circular scars when pulled away from the twig.
The female cones are borne upright on the branches and disintegrate
when mature. Many cultivars of fir are grafts of side branches,
which may begin as low, spreading plants but often later grow
upright.

The short key below should aid in identifying four of the more
common firs. It is followed by a longer, more detailed key that
includes those and 19 other, less common species.

Key to 4 Common Species

47. *A. homolepis*

1. Twigs deeply grooved and ridged, without hairs
   (Fig. 47) . . . . *A. homolepis*
   Nikko fir

1. Twigs smooth, with or without hairs (Fig. 48).
   2. Leaves silvery, glaucous, mostly longer than 4 cm (1½ in);
      twigs lacking hairs . . . . *A. concolor*
      White fir

48. *A. balsamea*

   2. Leaves dark green, less than 2.5 cm (1 in) long; twigs hairy.
      3. Twigs with dark gray hairs; cone bracts hidden. . .*A. balsamea*⁺
         Balsam fir

      3. Twigs with red hairs (may appear gray because of dust
         particles on the hairs); cone bracts protruding . . . . *A. fraseri*
         Fraser fir

Key to 23 Species

49. *A. grandis*

1. Twigs lacking hairs, smooth (Fig. 49).
   2. Leaves with white longitudinal bands or white spots on upper
      and lower surface.
      3. White longitudinal bands or white spots only near the
         tip of upper leaf surface (Fig. 50).
         4. Twigs deeply grooved (Fig. 51).

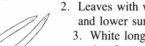

50. *A. cephalonica*

            5. Leaf edges rolled under, leaves less than 3.5 cm (1⅓
               in) long.
               6. Buds 3.5 cm (1⅓ in) long; tips of leaves on young
                  plants often divided into 2 sharp points . . . . *A. firma*
                  Momi fir

               6. Buds 2 cm (¾ in) long; tips of leaves on young
                  plants with single sharp point . . . . *A. chensiensis*
                  Ernest fir

            5. Leaf edges flat, leaves to 4.5 cm (1¾ in)

51. *A. chensiensis*

               long . . . . *A. holophylla*
               Needle fir

         4. Twigs smooth or with shallow grooves (Fig. 52).
            7. Buds not resinous or sticky; leaves pointed forward
               toward the twig tip, nearly parallel to the twig
               axis . . . . *A. cilicica*
               Cilician fir

            7. Buds resinous or sticky; leaves spreading from

52. *A. holophylla*

               twig axis more at right angles.
               8. Twigs with stripes or lines (Fig. 53); leaves with
                  sharp, often membranous tips.
                  9. Leaves less than 2.5 cm (1 in) long . . . .*A. cephalonica*
                     Greek fir

53. *A. cephalonica*

54. *A. grandis*

55. *A. magnifica*

56. *A. homolepis*

57. *A. cephalonica*

58. *A. balsamea*

9. Leaves greater than 2.5 cm (1 in) long ......*A. holophylla*
   Needle fir
8. Twigs smooth (Fig. 54), olive green to light brown;
   leaves rounded or notched at the tip.
   10. Leaves stout, 1–2 cm (⅓–¾ in) long ....*A. numidica*
      Algerian fir
   10. Leaves more flexible, 3.5–5.5 cm (1¼–2¼ in)
      long .... *A. grandis*
      Grand fir
3. White longitudinal bands along most of upper leaf surface
   (Fig. 55).
   11. Leaves greater than 4 cm (1½ in) long, silver,
      glaucous .... *A. concolor*
      White fir
   11. Leaves less than 4 cm (1½ in) long.
      12. Leaves glaucous, silver.
         13. Leaves 1–2 cm (⅓–¾ in) long, rigid, spreading
            at right angles from the twig, very
            sharp-pointed .... *A. pinsapo*
            Spanish fir
         13. Leaves 2.0–3.5 cm (¾–1⅜ in) long, curved at
            base like a hockey stick .... *A. magnifica*
            Red fir
      12. Leaves dark green .... *A.* ×*bornmuelleriana*
         Bornmueller fir
2. Leaves with white longitudinal bands on lower surface
   only.
   14. Twigs deeply grooved (Fig. 56) ....*A. homolepis*
      Nikko fir
   14. Twigs shallowly grooved, with or without stripes or
      lines (Fig. 57).
      15. Leaves less than 2 cm long (¾ in), usually with blunt or
         notched tips; twigs grooved and ridged.
         16. Buds 5 mm (⅛ in) or shorter; leaves very white on
            lower surface .... *A. koreana*
            Korean fir
         16. Buds usually longer than 5 mm (⅛ in); leaves
            spreading outward or even upward .... *A. numidica*
            Algerian fir
      15. Leaves mostly more than 2 cm (¾ in) long, with
         sharp, often membranous tips; twigs with grooves and
         lines or stripes .... *A. cephalonica*
         Greek fir
1. Twigs hairy (Fig. 58).
   17. Leaves with white longitudinal bands on both upper and
      lower surface.

59. *A. fraseri*

60. *A. holophylla*

61. *A. balsamea*

62. *A. magnifica*

63. *A. lasiocarpa*

64. *A. lasiocarpa*

65. *A. magnifica*

66. *A. balsamea*

18. Leaves with white longitudinal bands or white spots only near tip of upper surface (Fig. 59).
  19. Twigs grooved (Fig. 60).
    20. Leaf edges flat; leaves to 4.5 cm (1¾ in) long ... *A. holophylla*
          Needle fir
    20. Leaf edges rolled under; leaves less than 3.5 cm (1⅓ in) long.
      21. Buds 3.5 cm (1⅓ in) long; tips of leaves on young plants often divided into 2 sharp points .... *A. firma*
          Momi fir
      21. Buds 2 cm (¾ in) long; tips of leaves on young plants with 1 sharp point .... *A. chensiensis*
          Ernest fir
  19. Twigs smooth (lacking grooves) or with shallow grooves (Fig. 61).
    22. Buds resinous or sticky.
      23. Twigs lightly grooved; rare .... *A. amabilis*
          Pacific silver fir
      23. Twigs smooth (lacking grooves); common.
        24. Twigs with red hairs (may appear gray or black because dust particles cling to the hairs); cone bracts protruding .... *A. fraseri*
          Fraser fir
        24. Twigs with gray hairs; cone bracts hidden .... *A. balsamea*⁺
          Balsam fir
    22. Buds dry (not resinous, not sticky).
      25. Twigs yellow-green; leaves without white on midrib .... *A. cilicica*
          Cilician fir
      25. Twigs light brown to reddish brown; leaves sometimes with white on the midrib of lower surface ... *A. nephrolepis*
          Kinghan fir
18. Leaves with white longitudinal bands along most of upper surface (Fig. 62).
  26. Upper leaf surface deeply grooved or grooved partway and then ridged .... *A. procera*
          Noble fir
  26. Upper leaf surface flat or only shallowly grooved.
    27. Twigs lightly grooved, with ridges below the leaf bases (Fig. 63).
      28. Leaves twisted at base (Fig. 64) .... *A. lasiocarpa*
          Subalpine fir
      28. Leaves curved at base like a hockey stick (Fig. 65) .... *A. magnifica*
          Red fir
    27. Twigs smooth (lacking grooves) (Fig. 66).
      29. Twigs with red hairs (may appear gray or black

because dust particles cling to the hairs); cone bracts
protruding . . . . *A. fraseri*
Fraser fir

29. Twigs with gray hairs; cone bracts hidden . . . . *A. balsamea*[+]
Balsam fir

17. Leaves with white longitudinal bands on lower surface only.

30. Buds dry (not sticky, not resinous).

31. Leaves sometimes with white on midrib of lower surface, the edges strongly rolled under, less than 2 mm (1/16 in) wide . . . . *A. nephrolepis*
Kinghan fir

31. Leaves lacking white on midrib, edges only slightly rolled under; often 2 mm (1/16 in) wide or wider.

32. Twigs brown to gray, with minute lines or stripes; leaves in several planes, usually more than 2 mm (1/16 in) wide and less than 3 cm (1 1/4 in) long.

33. Leaf scars broadly elliptical; bud scales rounded at apex . . . . *A. alba*
European silver fir

33. Leaf scars circular; bud scales blunt or with dull point at apex, minutely toothed on the edges . . . . *A. nordmanniana*
Nordmann fir

32. Twigs brown-green to yellow, without lines or stripes; leaves in roughly one plane, nearly parallel to the twig axis, usually less than 2 mm (1/16 in) wide and often to 4 cm (1 1/2 in) long . . . . *A. cilicica*
Cilician fir

30. Buds resinous (may be sticky, crusty, or white).

34. Twigs grooved or ridged or with lines (Fig. 67).

67. *A. holophylla*

35. Leaf edges flat, not rolled under; leaves to 4.5 cm (1 3/4 in) long . . . . *A. holophylla*
Needle fir

35. Leaf edges rolled under, if only slightly; leaves less than 3.5 cm (1 1/3 in) long.

36. Buds 3.5 cm (1 1/3 in) long, with many scales; leaf tips on young plants divided into 2 sharp points . . . . *A. firma*
Momi fir

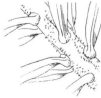

68. *A. sachalinensis*

36. Buds 2 cm (3/4 in) long or less; leaf tips rounded or with 1 point.

37. Twigs densely hairy (Fig. 68); buds hairy; leaves only 1.5 mm (1/16 in) wide at the middle and near the tips . . . . *A. sachalinensis*
Sakhalin fir

37. Twigs with widely scattered hairs (Fig. 69); buds smooth; leaves more than 1.5 mm (1/16 in) wide.

69. *A. chensiensis*

38. Leaves more than 2 mm wide at the middle and even wider toward the tips, with green upper surface and very white lower surface . . . . *A. koreana*
Korean fir

38. Leaves less than 2 mm wide throughout, with dark green upper surface and white lower surface . . . . *A. chensiensis*
Ernest fir

70. *A. balsamea*

34. Twigs smooth (not grooved, not ridged, and without lines) (Fig. 70).

39. Leaves mostly more than 3 cm (1¼ in) long; twigs sparsely hairy . . . . *A. grandis*
Grand fir

39. Leaves mostly less than 3 cm (1¼ in) long; twigs densely hairy.

40. Leaves tapered and then abruptly enlarged at the base . . . . *A. veitchii*
Veitch fir

40. Leaves not significantly enlarged at the base.

41. Twigs with red hairs (may appear gray or black because of dust particles on the hairs); cone bracts protruding . . . . *A. fraseri*
Fraser fir

41. Twigs with gray hairs; cone bracts hidden . . . . *A. balsamea*⁺
Balsam fir

1.   *A. alba* **Mill.   European silver fir.**

Native to the mountainous regions of Europe. A very hardy fir that grows slowly in North America. Unusual to find a mature specimen that has not lost many lower branches. Leaves 1.5–3.3 cm (¾–1¼ in) long. Cones 10–14 cm (4–5½ in) long.

'Aurea'   Leaves yellow.

'Brevifolia'   Leaves short.

'Columnaris'   Growth columnar; branches short, numerous; leaves slender, short.

'Compacta'   Growth dwarf, columnar, dense; branches erect.

'Elegans'   Growth dwarf; leaves short.

'Fastigata'   Growth columnar.

'Globosa'   Growth rounded.

'Green Spiral'   Trunk curved; branches short, bent backward; leaves crowded.

'Microphylla'   Growth dwarf; branches thin, short; leaves thin.

'Nana'   Growth slow, rounded.

'Pendula'   Branches drooping.

'Pyramidalis'   Growth conical; branches erect.

'Tenuiorifolia'   Leaves long, thin; cones longer.

'Tortuosa'   Branches stunted, twisted; leaves bent, irregularly arranged.

'Virgata'   Branches long, slender; twigs crowded near branch tips.

2.   *A. amabilis* Forbes.   Pacific silver fir.   Cascade fir.

Native to northwestern North America. Rare and difficult to grow in the Northeast. Leaves 2.5 cm (1 in) long. Cones 9–15 cm (3½–6 in) long.

'Compacta'   Growth dwarf, dense.

'Spreading Star'   Growth low, spreading.

+3.   *A. balsamea* (L.) Mill.   Balsam fir.

The only fir native to northeastern North America. Grows at higher altitudes and in swamps. Makes a good ornamental when young but is hurt by late frosts. Loss of branches as the tree ages, giving older specimens a ragged appearance. Foliage dark green with strong aroma, making the species a premium Christmas tree. Leaves 1.5–2.5 cm (⅗–1 in) long, with two white longitudinal bands on lower surface and some or no white on the upper surface. Cones 4–6 cm (1½–2½ in) long.   var. *phanerolepis* Fern. Cones smaller, with bracts protruding.

'Andover'   Growth spreading over the ground.

'Argentea'   Leaves with white tips.

'Globosa'   Growth rounded; similar to 'Nana'.

'Hudsonia'   Growth dwarf, spreading.

'Nana'   Shrub; growth slow, rounded, dense; twigs pointing forward at 45 degree angle; leaves very short, narrow.

'Quintin's Spreader'

'Variegata'   Leaves variegated white.

'Verkade's Prostrate'   Growth spreading over the ground.

4. *A. × bornmuelleriana* **Mattf.** (*A. cephalonica* **Loud.** × *A. nordmanniana* **Spach.)** **Bornmueller fir.** **Turkey fir.**

> Native to Turkey. A natural cross between *A. cephalonica* and *A. nordmanniana*. Leaves 2.5–4.0 cm (1–1½ in) long. Cones 12–15 cm (4¾–6 in) long.

5. *A. cephalonica* **Loud.** **Greek fir.**

> Native to Greece. Handsome dark green tree and hardy. Leaves with sharp-pointed pale tips, 1.5–2.5 cm (⅝–1 in) long. Cones 13–18 cm (5–7 in) long.

> 'Aurea'   Young leaves yellow.
> 'Meyer's Dwarf'   Growth dwarf, spreading; leaves 8–15 mm (¼–⅝ in) long.
> 'Robusta'   Branches strong; twigs thick, stiff; leaves crowded; cones larger.
> 'Rubiginosa'   Foliage at ends of branches sometimes red-brown.
> 'Submutica'   Cones smaller.

6. *A. chensiensis* **Van Tiegh var.** *ernestii* **(Rehd.) Liu.** **Ernest fir.** **Ernst fir.**

> Native to China. Leaves 1–3 cm (⅜–1⅛ in) long. Cones 5–10 cm (2–4 in) long.

7. *A. cilicica* **(Ant. & Kotschy) Carr.** **Cilician fir.**

> Native to Asia. Leaves 2.5–4.0 cm (1–1½ in) long. Cones 16–30 cm (6–12 in) long.

8. *A. concolor* **(Gord.) Hoopes.** **Concolor fir.** **White fir.** **Colorado fir.**

> A beautiful fir from the forests of western North America with long, glaucous, blue-gray leaves. Popular as an ornamental and now being

planted to some extent for Christmas trees. Can be extremely slow growing and does not thrive in compacted soils. Leaves 4–5 cm (1½–2 in) long. Cones 8–13 cm (3–5 in) long.   var. *lowiana* (Gord.) Lemm. Buds smaller; leaves with less gray on upper surface.

'Albospica'   Young leaves white.

'Archer's Dwarf'   Growth dwarf, wider than high; twigs curling up.

'Argentea'   Leaves white.

'Aurea'   Leaves yellow, turning green.

'Brevifolia'   Leaves short, thick, blunt.

'Candicans'   Leaves blue to nearly white.

'Clarence'

'Compacta'   Growth dwarf, compact, irregular; leaves narrow, glaucous.

'Conica'   Growth slow, columnar.

'Elkins Weeping'

'Fagerhult'   Growth slow; leaves long.

'Fastigata'   Growth columnar; branches short, upright.

'Gable's Weeping'   Growth dwarf, spreading; branches drooping.

'Globosa'   Growth dwarf, rounded.

'Green Globe'   Growth dwarf, rounded, becoming conical.

'Masonic Broom'   Growth dwarf, rounded.

'Pendula'   Growth columnar, with sharply drooping branches.

'Piggelmee'   Leaves long, crowded.

'Pineola Dwarf'   Growth dwarf.

'Pyramidalis'   Growth upright, vigorous, but branches angling down.

'Schrammii'   Similar to 'Violacea'; rare in United States.

'Select Blue'

'Verkades Witchbroom'   Growth slow.

'Violacea'   Leaves blue; rare.

'Wattezii'   Growth low, spreading over the ground; leaves white.

'Wintergold'   Leaves yellow, then flecked green.

9.   *A. firma* Sieb. & Zucc.   Momi fir.

Native to Japan. Leaves divided into two sharp points on young plants, blunt or notched on older plants, 1.5–3.5 cm (⅝–1½ in) long. Cones 8–15 cm (3¼–6 in) long.

10. *A. fraseri* (Pursh) Poir. Fraser fir. Southern balsam fir. She-balsam.

A prized Christmas tree. Native to the mountains of southeastern United States. Closely related to balsam fir. Can make a nice ornamental but like balsam may lose its branches with age. Leaves 1.5–2.5 cm (⅗–1 in) long, usually whiter than balsam fir on lower surface. Each band composed of 8–12 dashed lines. Cones 4–6 cm (1½–2¼ in) long, with protruding bracts.

'Klein's Nest'
'Pendula'   Branches drooping.
'Prostrata'   Growth low, spreading over the ground.

11. *A. grandis* (D. Don ex Lamb.) Lindl.   Grand fir.

Native to western North America. Leaves thin, flexible, 1.3–4.5 cm (½–1¾ in) long. Cones 7–12 cm (2¾–4¾ in) long.

'Nana'   Growth dwarf.

12. *A. holophylla* Maxim.   Needle fir.

Native to Manchuria and Korea. Leaves 2.0–4.5 cm (¾–1¾ in) long. Cones 9–14 cm (3½–5½ in) long.

13. *A. homolepis* Sieb. & Zucc.   Nikko fir.   Niko fir.

Native to Japan. Grows well in northeastern North America. Leaves dark green but will persist in a yellowed condition for some time if growing conditions are not close to excellent. Twigs and branches deeply grooved. Leaves 2.5 cm (1 in) long. Cones 7–12 cm (2¾–4¾ in) long.   var. *umbellata* (Mayr.) Wils. Cones green-yellow with flattened apex and raised center.

'Prostrata'   Growth spreading; foliage dark green.
'Scottae'   Growth dwarf.
'Tomonii'   Growth dwarf; branches slender, sparse, leaves short.

14.   *A. koreana* **Wils.   Korean fir.**

Native to Korea. Leaves with two brilliant white longitudinal bands
on lower surface, 1–2 cm (⅜–¾ in) long. Cones 4–7 cm (1⅝–2¾
in) long.

'Aurea'   Leaves yellow.
'Blauer Pfiff'   Leaves blue-green.
'Blue Standard'   Cones dark purple.
'Compact Dwarf'   Growth dwarf, compact.
'Horstmann's Silver Lock'   Growth rapid; leaves curved back, silvery
    on lower surface.
'Nisbet'   Growth dwarf.
'Piccolo'   Leaves short, divided at tips.
'Prostrata'   Growth low, spreading over the ground.
'Prostrate Beauty'   Growth spreading over the ground.
'Silver Show'   Leaves turned upward so that the white lower surface
    shows.
'Starker's Dwarf'   Growth slow, rounded, dense, flat-topped.

15.   *A. lasiocarpa* **(Hook.) Nutt.   Subalpine fir.**

Native to northwestern North America. Leaves blue-green, glaucous, 2.5–4.0 cm (1–1½ in) long. Cones 8–10 cm (3–4 in) long.
var. *arizonica* (Merr.) Lemm. Corkbark fir. Bark thick, corky;
cones 7–8 cm (2¾–3¼ in) long.

'Argentea'   Leaves glaucous.
'Beissneri'   Growth dwarf; branches and leaves twisted.
'Compacta'   Growth dwarf, rounded, compact.
'Conica'   Growth dwarf, conical, dense.
'DuFlon'   Growth dwarf.
'Glauca'   Leaves glaucous.
'Mulligan's Dwarf'   Growth slow.
'Pendula'   Branches drooping.
'Roger Watson'   Growth dwarf, dense.

16.   *A. magnifica* **A. Murr.   Red fir.**

Native to northwestern North America. Rarely cultivated and, like
many Pacific northwestern conifers, does not grow well in the

Northeast. Leaves curved at base like a hockey stick, 2.0–3.5 cm (¾–1⅜ in) long. Cones 11–22 cm (4¼–8½ in) long.

17. *A. nephrolepis* (Trautv.) Maxim.   Kinghan fir.

Native to China, Korea, and Siberia. Rarely cultivated. Leaves 1–3 cm (½–1¼ in) long. Cones 4.5–7.5 cm (1¾–3 in) long.

18. *A. nordmanniana* (Steven) Spach.   Nordmann fir. Caucasian fir.

Native to Greece, Caucasus, and Asia. Leaves 2–3 cm (¾–1¼ in) long. Cones 10.5–16.5 cm (4–6½ in) long.

'Aurea' Leaves yellow.
'Aureovariegata' Some leaves partly or all yellow.
'Compacta' Growth rounded; branches short, crowded.
'Glauca' Leaves glaucous.
'Golden Spreader' Growth low, spreading; leaves yellow.
'Horizontalis' Growth dwarf.
'Jensen' Branches drooping.
'Nana' Growth dwarf.
'Pendula' Branches drooping.
'Procumbens' Growth slow, spreading over the ground.
'Prostrata' Growth low, spreading.
'Refracta' Growth rounded; leaves turned upward.
'Robusta' Branches thick; leaves crowded.
'Tortifolia' Some leaves twisted, turning up.

19. *A. numidica* de Lann.   Algerian fir.

Native to Algeria. Leaves 1–2 cm (⅜–¾ in) long. Cones 12–18 cm (4¾–7 in) long.

'Glauca' Leaves shorter and wider.
'Lawrenceville' Growth conical; foliage dark green.
'Pendula' Branches crowded, drooping.

---

20.   *A. pinsapo* Boiss.   Spanish fir.

Native to Spain. Leaves rigid, thick, 1–2 cm (⅜–¾ in) long. Cones 10–15 cm (4–6 in) long.

'Aurea'   Growth slow; new growth yellow.
'Clarke'   Growth dwarf.
'Fastigiata'   Growth columnar; branches short.
'Glauca'   Growth slow and spreading when young, becoming upright; leaves more glaucous.
'Horstmann'   Growth dwarf; branches crowded; leaves short.
'Kelleriis'   Branch tips turning backward.
'Nana'   Growth dwarf.
'Pendula'   Branches drooping.

21.   *A. procera* Rehd.   Noble fir.

Native to northwestern North America. Leaves 2.5–3.5 cm (1–1¼ in) long. Cones 10–15 cm (4–6 in) long.

'Aurea'   Foliage yellow.
'Blauehexe'   Growth rounded; leaves short, crowded.
'Glauca'   Growth slow, compact, spreading when young, becoming upright; leaves more glaucous.
'Glauca Prostrata'   Growth spreading over the ground, flat-topped; foliage blue.
'Jeddeloh'   Growth slow, rounded.
'Nobel'   Leaves crowded, curved downward.
'Prostrata'   Growth dwarf, spreading over the ground.
'Sherwoodii'   Leaves yellow.

22.   *A. sachalinensis* (Fr. Schm.) Mast.   Sakhalin fir.   Sachalin fir.

Native to northern Japan and Soviet Union (Sakhalin Island). Leaves 1.5–3.0 cm (⅝–1¼ in) long. Cones 5–8 cm (2–3 in) long.

23.   *A. veitchii* Lindl.   Veitch fir.

Native to Japan. Leaves 1.0–2.5 cm (⅜–1 in) long. Cones 4–6 cm (1½–2½ in) long.

Calocedrus decurrens

Calocedrus Kurz.   Incense-cedar.

This genus includes two species from Asia and one species from western North America. These species, which were formerly placed in the genus *Libocedrus*, are probably now more correctly considered members of *Heyderia*, but for the present it is convenient to retain the use of *Calocedrus*.

1.   *C. decurrens* (Torr.) Florin.   Incense-cedar.

A native tree of the northwestern United States. In overall appearance similar to *Chamaecyparis* and *Thuja*. A pleasing specimen in cultivation, although a protected location is usually necessary. Twigs flattened. Leaves aromatic, bright green, scalelike with glands on their backs, joining the twig for some distance along the twig, 4 in a whorl around the twig. Cones 2.0–2.5 cm (¾–1 in) long.

'Aureovariegata'   Foliage variegated yellow.

'Columnaris'   Growth columnar; branches short.

'Compacta'   Growth compact, dense.

'Depressa'   Growth rounded, dense, compact, as wide as high.

'Glauca'   Leaves glaucous.

'Greenspire'   Growth columnar; leaves dark green.

'Horizontalis'   Branches spreading.

'Intricata'   Growth dwarf; branches thick; foliage somewhat brownish in winter.

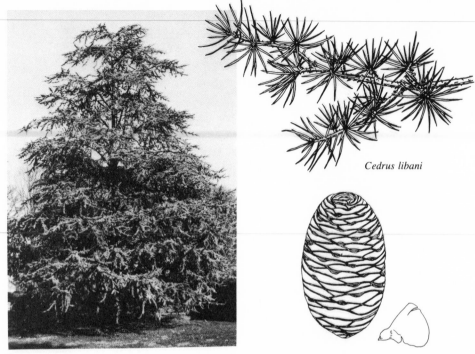

Cedrus libani

Cedrus atlantica

Cedrus Trew.   Cedar.

This is a genus of stout trees with stiff, irregularly spaced branches. The four species are native to Asia and Africa. Only three of these are cultivated to any extent. The leaves are clustered in groups of about 8–20 but are solitary on young shoots. Cones are upright and disintegrate at maturity. Members of this genus are sporadically hardy farther north than southern Pennsylvania and Long Island such as along the Atlantic Coast and in the warmer climates created by the Great Lakes.

1. Leading shoot upright, stiff; branch tips rarely drooping.
    2. Twigs bearing dense, short hairs; leaves as high or higher than wide, often blue .... *C. atlantica*
       Atlas cedar
    2. Twigs lacking hairs or only sparsely hairy; leaves as wide or wider than high, light green .... *C. libani*
       Cedar-of-Lebanon
1. Leading shoot and branch tips hanging or drooping... *C. deodara*
   Deodar cedar

1. *C. atlantica* (Endl.) G. Manetti ex Carr.   Atlas cedar.   Blue Atlas cedar.

Native to North Africa. One of the more spectacular ornamental conifers if the location is coastal or a protected area in the southern part of the Northeast. Leaves blue, gray, or silver, usually less than 2.5 cm (1 in) long. Cones 5–8 cm (2–3 in) long.

'Albospica'   Young leaves nearly white.

'Argentea'   Leaves silver.

'Argentea Fastigiata'   Growth columnar; foliage blue.

'Aurea'   Young leaves yellow.

'Aurea Robusta'   Growth conical; leaves yellow.

'Fastigiata'   Growth columnar; branches short, thick.

'Glauca'   Leaves glaucous; branches close, thick; hardier.

'Glauca Fastigiata'   Growth columnar.

'Glauca Horizontalis'   Branches spreading; leaves brilliant blue.

'Glauca Pendula'   Main trunk and branches drooping.

'Pendula'   Main stem and branches drooping.

'Pyramidalis'   Growth columnar or conical.

'Rustic'   Leaves very blue.

'Variegata'   Young leaves pale yellow.

'Wilkman'   Foliage rich green.

2. *C. deodara* (Roxb. ex Lamb.) G. Don in Loud.   Deodar cedar.

Native to the Himalayas. Softer in appearance than the other species. Leaves green to blue-green, 2.5–5.0 cm (1–2 in) long. Cones 8–10 cm (3–5 in) long, 5–9 cm (2–3½ in) wide.

'Albospica'   Growth conical; new foliage white.

'Argentea'   Leaves silver.

'Aurea'   Leaves yellow.

'Aurea Pendula'   Branches drooping; leaves yellow.

'Aurea Wells'   Same as 'Wells Golden'.

'Compacta'   Growth rounded, dense.

'Crassifolia'   Tree stiff, stunted; leaves short, thick.

'Cream Puff'   Growth slow; foliage dense, with white tones.

'Descano Dwarf'   Dwarf, compact, semiweeping.

'Eisregen' Very hardy in Europe.

'Eiswinter' Hardy.

'Erecta' Leaves light green to white.

'Glauca' Leaves glaucous.

'Gold Cone' Growth conical; branch tips drooping.

'Gold Rush' Leaves yellow.

'Gold Strike' Growth columnar; leaves yellow.

'Golden Horizon' Leaves short, yellow on exposed sites.

'Hesse' Growth dwarf.

'Hibernal'

'Karl Fuchs' Leaves blue.

'Kashmir' Leaves blue; hardy.

'Kingsville'

'Klondike' Growth conical, almost as wide as high; foliage with
    yellow tint.

'Nana' Growth dwarf.

'Nivea' Growth bushy, wide; young leaves nearly white.

'Paktia' Leaves tending toward white.

'Pendula' Stem drooping; branches long and drooping.

'Polarwinter' Supposedly hardy.

'Prostrata' Growth slow.

'Pygmea' Growth dwarf and rounded.

'Raraflora Gold Prostrate'

'Repandens' Branches strongly drooping.

'Repens' Growth low and spreading.

'Robusta' Branches stout; leaves rigid.

'Silver Mist' Growth dwarf, mound-shaped.

'Snow Sprite' Growth dwarf, mound-shaped.

'Tristis' Branches short.

'Verticillata' Growth compact; leaves glaucous.

'Verticillata Glauca' Growth conical, horizontal; branches few; leaves
    glaucous.

'Viridis' Leaves shining, deep green.

'Viridis Prostrata' Growth spreading over the ground; leaves
    shining deep green.

'Wells Golden' Foliage yellow.

'White Imp' Growth dwarf; foliage white; similar to 'Snow Sprite'.

'Wiesemannii' Growth upright; branches crowded; leaves crowded.

3. *C. libani* A. Rich.   Cedar-of-Lebanon.

Native to Asia and the most hardy species of the genus. Leaves light green, at least 2.5 cm (1 in) long. Cones 8–10 cm (3–4 in) long, 5 cm (2 in) wide.

'Argentea' Foliage silver.

'Aurea' Growth weak; leaves light green to yellow.

'Aurea Prostrata' Growth low, flat, spreading; leaves light green to yellow.

'Comte de Dijon' Growth dwarf, conical, compact, dense.

'Conica Nana' Growth dwarf; branches turning up.

'Decidua' Growth bushy; branches short and crowded.

'Denudata' Branches very irregularly spaced.

'Glauca' Leaves glaucous.

'Glauca Pendula' Branches drooping; leaves glaucous.

'Gold Tip' Shrub; growth irregular; leaves yellow to green.

'Golden Dwarf' Same as 'Aurea Prostrata'.

'Green Knight'

'Multicaulis' Branches drooping, numerous; twigs short.

'Nana' Growth dwarf, compact; leaves thin, short.

'Nana Pyramidata' Growth dwarf; young shoots stiff, thin.

'Pendula' Branches drooping.

'Sargentii' Growth dwarf; leaves long, thick, blue.

'Stricta' Branches short, crowded.

'Tortuosa' Branches and twigs twisted.

'Viridis' Leaves shining green.

Cephalotaxus harringtonia var. *drupacea*

Cephalotaxus Sieb & Zucc.   Plumyew.

This genus of small trees or, more often, branching shrubs contains nine species, all native to Asia. New York and protected coastal areas are the northern hardiness limits for cultivated plumyews. A berrylike or plumlike female reproductive structure separates *Cephalotaxus* from most other genera of conifers. The purple color of this structure, together with the opposite leaves and opposite or whorled branches, distinguishes the genus from *Taxus*. The presence of a stalk bearing the reproductive structure and a prominent midrib on the leaves differentiate this genus from *Torreya*.

71. *C. harringtonia*

72. *C. fortunii*

1. Leaves abruptly pointed (Fig. 71), 2.0–4.5 cm
   (¾–1¾ in) long . . . .*C. harringtonia*
                                                                    Japanese plumyew
1. Leaves gradually tapering to a point (Fig. 72), 4.5–6.5 cm
   (1¾–2½ in) long . . . . *C. fortunii*
                                                                    Chinese plumyew

1. *C. fortunii* **Hook.   Chinese plumyew.**

Shrub with dark green, lustrous foliage. Native to China. Leaves 4.5–6.5 cm (1¾–2½ in) long. Female reproductive structure 1–2 cm (⅜–¾ in) wide and 2.0–3.5 cm (¾–1⅜ in) long.

'Alpina'   Leaves shorter; male cones without stalks.
'Brevifolia'   Leaves shorter.
'Concolor'   Growth dwarf.
'Grandis'   Leaves longer.
'Longifolia'   Leaves longer.
'Pendula'   Branches drooping.
'Robusta'   Leaves flatter, longer.

2. *C. harringtonia* **(D. Don) C. Koch.   Japanese plumyew.**

Shrub, native to Japan. Uncommon but more likely to be found in cultivation than the preceding species. Leaves 2.0–4.5 cm (¾–1¾ in) long. Female reproductive structure 1.3–2.5 cm (½–1 in) wide and 2–3 cm (¾–1¼ in) long.   var. *drupacea* (Sieb. & Zucc.) Koidz. Branches never drooping; leaves close together, two-ranked.

'Fastigata'   Growth columnar; leaves spirally arranged.
'Gnome'   Growth dwarf.
'Koreana'   Foliage tufted.
'Nana'   Growth slow, upright.
'Prostrata'   Growth low, spreading.
'Sphaeralis'   Seeds spherical.

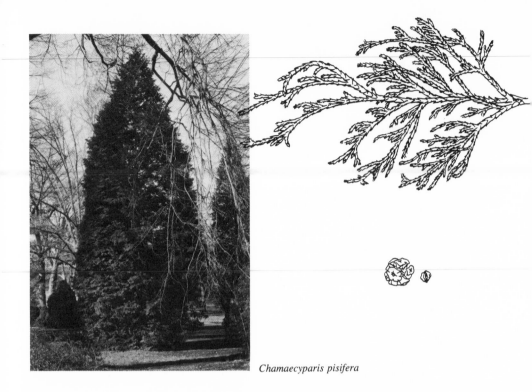

Chamaecyparis pisifera

Chamaecyparis Spach.   False-cypress.

Chamaecyparis is a genus of six species in Asia and North America, with one species native to northeastern North America. The many horticultural varieties are used frequently in specimen plantings as trees or shrubs. Some cultivars of *C. lawsoniana* and *C. pisifera* bear only awl-shaped or needlelike leaves. Most, however, display the typical scalelike leaves, often with white or yellow variegation. The usually flattened sprays are flexible, giving large trees a graceful appearance. The genus can be difficult to distinguish from *Thuja* and *Platycladus* unless cones are present. The lateral leaves of *Chamaecyparis* in most cases meet on a visible seam, whereas in *Thuja* they are spread apart with no seam visible. In the former the leaves are also smaller, the differences being slight (0.5 mm × 1 mm versus 1 mm × 2 mm). *Chamaecyparis* has a prominent groove on both facial and lateral scalelike leaves, while in *Platycladus* usually just the facial leaves display a shallow groove, pit, or no indentation at all. *Chamaecyparis* is distinguished from *Cupressus* by its flattened sprays, smooth-edged leaves, and greater number of seeds per cone scale and from *Thuja* by its spherical cones with shield-shaped scales. The key below refers to scalelike leaves on

73. *C. thyoides*

74. *C. obtusa*

75. *C. lawsoniana*

76. *C. nootkatensis*

77. *C. nootkatensis*

78. *C. pisifera*

first-year shoots. (For definitions and illustrations of the terms *facial leaves, gland,* and *lateral leaves,* see p. 16.)

1. Leaves all needlelike or awl-shaped .... some cultivars of
   C. lawsoniana and
   C. pisifera
1. Leaves all or mostly scalelike.
   2. Seam where lateral leaves meet usually visible; leaf tips blunt, or at least not sharply pointed, pressed close to the twig; leaves not sharply keeled (Fig. 73).
      3. Visible portion of lateral leaves at least 2 times as long as that of the facial leaf (Fig. 74); twigs (including leaves) 2 mm (¹⁄₁₆ in) in diameter .... *C. obtusa*
         Hinoki-cypress
      3. Visible portion of lateral leaves about equal in length to or slightly longer than that of the facial leaf (Fig. 75); twigs (including leaves) 1 mm (¹⁄₃₂ in) in diameter.
         4. Leaves with white markings on lower surface of spray, often bluish; male cones red .... *C. lawsoniana*
            Port-Orford-cedar
         4. Leaves lacking white markings on lower surface, green, clustered at the end of the branch; male cones yellow .... *C. thyoides*⁺
            White-cedar
   2. Seam where lateral leaves meet usually not visible; leaf tips sharply pointed, spreading away from the twig; leaves sharply keeled (Fig. 76).
      5. Twigs appearing 4-angled; leaves sharply keeled, lacking white markings; visible portion of lateral leaves tending to be equal in length to or longer than that of the facial leaf (Fig. 77).
         6. Cone scales 4–6, each bearing 2 seeds; foliage odorous (like cut potatoes) when bruised .... *C. nootkatensis*
            Alaska-cedar
         6. Cone scales 8, each bearing 5 seeds; foliage less odorous when bruised .... see ×*Cupressocyparis leylandii*
            Leyland-cypress
      5. Twigs with more rounded appearance; leaves not as sharply keeled, with white markings on lower surface; visible portion of lateral leaves tending to be shorter than that of the facial leaf (Fig. 78) .... *C. pisifera*
         Sawara-cypress

1. *C. lawsoniana* (A. Murr.) Parl. Port-Orford-cedar. Lawson-cypress.

A large tree in its native North American West Coast habitat but in cultivation a smaller tree, often with glaucous or blue leaves. Male cones red. Female cones 1 cm in diameter, with 2–4 seeds. Cones 8 mm (¼ in) in diameter. Cone scales 8–10. Many, many cultivars produced in Europe, but few available in this country because of the climate and problems with growing nursery stock. Some cultivars described briefly below; European cultivars listed without description as follows:

'Alumii Magnifica'
'Argenteovariegata'
'Argenteovariegata Nova'
'Atrovirens'
'Aurea Nova'
'Aureospica'
'Aureovariegata'
'Beissneriana'
'Billwoodiana'
'Boeri'
'Casuarinifolia'
'Casuarinifolia
    Aureovariegata'
'Coerulea Erecta'
'Coerulescens'
'Compacta Nova'
'Cooperi'
'Crispa'
'Cristata'
'Darleyensis'
'Delorme'
'Depkenii'
'Dore de Croux'
'Epacroides'
'Erecta Aureospica'
'Erecta Glaucescens'
'Ericoides'
'Falcata'
'Filicifolia Pendula Nana'
'Filifera'
'Filiformis Glauca'
'Flavescens'
'Fletcher's White'

'Forsteckensis Variegata'
'Fragrans'
'Friesia'
'Glauca Globus'
'Glauca Lombartsii'
'Golden Prince'
'Golden Spire'
'Gracilis Aurea
    Pygmaea'
'Gracilis Glauca'
'Gracilis Nana'
'Gracilis Nova'
'Gracillima'
'Green Wall'
'Harkia'
'Intertexta Atrovirens'
'Juniperina'
'Kelleriis'
'Koosterhuis'
'Laxa'
'Luteogracilis'
'Magnifica Aurea'
'Melford'
'Milfordensis'
'Mimima
    Argenteovariegata'
'Moonlight'
'Nana Albovariegata'
'Nana Compacta'
'Nivea'
'Olbrichii'
'Overeynderi'
'Parsons'

'Pena Park'
'Pendula Alba'
'Pendula Aurea'
'Pendula Nova'
'Plumosa'
'Plumosa Glauca'
'Procumbens'
'Prostrata Glauca'
'Pulverulenta'
'Pyramidalis Alba
    Nana'
'Pyramidalis Lutea
    Gracilis'
'Raievskyana'
'Robusta Argentea'
'Robusta Aurea'
'Royal Gold'
'Shawii'
'Souvenir de Leide'
'Spek'
'Spiralis'
'Squarrosa'
'Stricta'
'Stricta Aurea'
'Stricta Excelsa'
'Stricta Glauca'
'Tilgate'
'Triomf van Lombarts'
'Van Eck'
'Van Tol'
'Weisseana'
'Westermannii
    Aureovariegata'

'Albospica'   Leaves at spray tips white.

'Albovariegata'   Shrub; growth conical; foliage patched white.

'Alumigold'   Growth conical; leaves yellow.

'Alumii'   Growth columnar, compact; twigs and sprays numerous.

'Alumii Nana Compacta'   Growth weakly conical.

'Argentea'   Leaves silvery gray on both sides.

'Ashton Gold'   Leaves yellow.

'Aurea'   Leaves yellow when young.

'Aurea Densa'   Growth dwarf; leaves yellow.

'Azurea'   Smaller tree.

'Blom'   Growth columnar, dense; twigs erect, vertical.

'Blue Gem'   Growth conical, compact; leaves blue.

'Blue Gown'   Growth conical; leaves glaucous blue.

'Blue Jacket'   Growth conical, broad, compact.

'Blue Nantais'   Growth conical, slow; foliage yellowish.

'Blue Plume'   Growth columnar, wide; similar to 'Plumosa Glauca'.

'Blue Surprise'   Growth conical; leaves needlelike.

'Booth'   Growth coarse.

'Bowleri'   Shrub; branches spreading, weak.

'Broomhill Gold'   Growth columnar; leaves yellow to yellow-green.

'Bruinii'   Growth conical; leaves glaucous.

'Caudata'   Shrub; foliage in dense, irregular tufts.

'Chilworth Silver'   Growth dwarf; leaves needlelike.

'Chingii'

'Coerulea'   Leaves blue-green.

'Columnaris'   Growth conical; twigs erect.

'Compacta'   Growth dwarf, conical, dense, compact.

'Croftway'   Growth conical; leaf tips white.

'Dow's Gem'   Branches drooping; leaves long.

'Drummondii'   Leaves small.

'Duncanii'   Shrub; twigs threadlike.

'Dwarf Blue'   Growth dwarf.

'Elegantissima'   Foliage yellow.

'Ellwoodii'   Growth conical and compact.

'Ellwoodii Glauca'   Growth columnar; leaves intense light blue.

'Ellwood's Gold'   Growth conical; leaf tips yellow.

'Ellwood's Pillar'   Branches crowded.

'Ellwood's Pygmy'   Growth dwarf.

'Ellwood's White'   Growth dwarf; foliage with white patches.

'Erecta'   Growth narrowly conical; leaves bright green.

'Erecta Alba'   Growth conical; twig tips white.

'Erecta Argenteovariegata'   Growth columnar; foliage patched white.

'Erecta Aurea'   Growth dwarf, conical; branches erect.

'Erecta Filiformis'   Growth conical; main branches erect, short and stiff; twigs threadlike.

'Erecta Filiformis Compacta'

'Erecta Glauca'   Growth columnar and slender.

'Erecta Viridis'   Growth columnar; branches dense, erect, vertical.

'Ericoides'   Growth dwarf; branches crowded; twigs stiff, long, thin.

'Erika'

'Felix'   Growth conical, robust.

'Filifera Glauca'   Growth conical; branches drooping.

'Filiformis'   Growth somewhat drooping, with threadlike sprays.

'Filiformis Compacta'   Growth dwarf, compact; twigs drooping.

'Fleckalwood'   Similar to 'Ellwood's White'.

'Fletcheri'   Growth columnar; twigs erect, reddish.

'Fletcheri Nana'   Growth dwarf.

'Fletcher's Compact'   Similar to 'Fletcheri Nana'.

'Fletcher's White'   Growth columnar; leaves speckled white.

'Forsteckensis'   Growth very dwarf; twigs short.

'Fraseri'   Growth conical and slender; twigs erect.

'Gimbornii'   Growth dwarf; sprays stiff, erect; leaves glaucous.

'Glauca'   Foliage glaucous.

'Glauca Argentea'   Growth conical; leaves blue-green, sometimes with white.

'Globosa'   Growth dwarf, rounded; branches and twigs erect; mature leaves may be awl-shaped.

'Gnome'   Growth dwarf, rounded to conical, similar to 'Forsteckensis'.

'Golden King'   Growth conical; leaves yellow.

'Golden Showers'   Growth compact; branches with drooping tips; leaves yellow.

'Golden Triumph'   Growth conical; branches erect; leaves yellow.

'Golden Wonder'   Growth conical; leaves yellow.

'Gold Splash'   Growth slow; foliage patched yellow.

'Gracilis'   Twig tips drooping.

'Gracilis Aurea'   Twigs drooping; leaves yellow.

'Gracilis Pendula'   Branches drooping.

'Grandi'   Growth slow; leaves glaucous.

'Grandis'   Growth dwarf, rounded; leaves dark, glaucous green.

'Grayswood Gold'   Foliage yellow.

'Grayswood Pillar'   Growth columnar.

'Green Globe'   Growth compact; foliage deep green.

'Green Hedger'   Growth conical, dense; leaves bright green.

'Green Pillar'   Leaves bright green, with golden tint.

'Green Spire'   Growth conical; foliage rich green.

'Hillieri'   Leaves gold; sprays light and feathery.

'Hogger' Growth conical.

'Hollandia' Growth conical; leaves dark green.

'Howarth's Gold' Growth conical; leaves light yellow.

'Imbricata Pendula'

'Intertexta' Branches weakly erect and drooping at ends.

'Intertexta Pendula' Branches drooping.

'Ivonne'

'Kelleriis Gold' Growth columnar, slender; leaves yellow.

'Kestonensis' Similar to 'Fletcheri'.

'Killiney Gold' Growth columnar; leaves yellow.

'Knowefeldensis' Growth dwarf.

'Krameri' Shrub; twigs long and threadlike.

'Lane' Growth columnar; twigs crowded and feathery; leaves yellow.

'Little Spire' Growth dwarf, conical, compact.

'Lombartsii' Growth conical; twigs stout; leaves yellow.

'Lutea' Growth columnar; outer leaves yellow.

'Lutea Nana' Growth dwarf, conical, dense; leaves yellow.

'Luteocompacta' Growth conical, dense; foliage yellow.

'Lutescens' Foliage yellow.

'Lycopodioides' Growth conical, dense; twigs twisted and spiral.

'Maas' Growth conical; leaves light green.

'Masonii' Growth conical, slower; hardy.

'Minima' Growth dwarf; leaves white on lower surfaces.

'Minima Aurea' Growth dwarf, compact; leaves yellow.

'Minima Glauca' Growth dwarf, rounded; leaves glaucous.

'Moerheimii' Growth conical; leaves yellow to yellow-green.

'Monumentalis' Twigs erect.

'Monumentalis Aurea' Growth conical, slender; spray tips yellow.

'Monumentalis Glauca' Growth dwarf; twigs erect; leaves speckled white.

'Monumentalis Nova' Growth columnar.

'Naberi' Growth conical; leaf color variously green and yellow.

'Nana' Growth dwarf but producing a leading shoot later.

'Nana Albospica' Growth dwarf, conical, dense; foliage white at twig tips.

'Nana Argentea' Growth dwarf, rounded to conical.

'Nana Argenteovariegata' Growth conical; foliage tipped white.

'Nana Glauca' Growth dwarf; foliage coarse.

'Nana Rogersii' Same as 'Rogersii'.

'Nestoides' Growth slow; outer leaves silvery; similar to 'Tamariscifolia'.

'New Golden'   Growth conical; leaves yellow.

'Nidiformis'   Growth dwarf; leaves long-pointed.

'Oregon Blue'   Growth conical; branch ends drooping.

'Patula'   Growth conical, compact.

'Pembury Blue'   Growth columnar; first-year leaves blue.

'Pendula'   Growth conical, drooping; leaves glossy, dark green.

'Pendula Vera'   Branches long and drooping.

'Pixie'   Growth rounded; branches numerous, crowded.

'Pottenii'   Growth columnar.

'President Roosevelt'   Growth conical; leaves yellow-spotted.

'Pygmaea'   Growth dwarf.

'Pygmaea Argentea'   Growth dwarf, rounded to conical.

'Pyramidalis'   Growth columnar, slender.

'Pyramidalis Alba'   Growth columnar; young leaves white.

'Rijnhof'   Growth rounded, often flat-topped; branches crowded; leaves
    needlelike.

'Robusta'   Growth columnar; leaves dark green.

'Robusta Glauca'   Growth columnar; sprays short and thick.

'Rogersii'   Growth dwarf; branches short and erect.

'Rosenthalii'   Growth conical.

'Schongariana'   Growth conical, dense.

'Silver Ball'

'Silver Gem'   Growth conical; young twigs with silver leaves.

'Silver Moon'   Growth slower, conical; branches spreading; leaves at
    twig tips silver.

'Silver Queen'   Growth conical; young leaves white.

'Silver Thread'   Growth dwarf, columnar.

'Silver Tip'

'Smithii'   Growth conical; leaves yellow.

'Snow Flurry'   Similar to 'Fletcher's White'.

'Snow Queen'

'Southern Gold'   Growth conical; leaves yellow, spotted with green.

'Stardust'   Growth conical; leaves yellow.

'Stewartii'   Growth conical; leaves yellow to light green.

'Tamariscifolia'   Shrub; twigs drooping, curved, and twisted.

'Tharandtensis'   Growth dwarf, rounded, compact.

'Tharandtensis Caesia'   Growth dwarf, rounded, compact.

'Tortuosa'   Growth conical; branches and twigs thick.

'Triomf van Boskoop'   Growth conical.

'Vens Yellow'

'Versicolor'   Growth conical; leaves spotted with white or yellow.
'Viner's Gold'   Growth columnar; foliage yellow.
'Wansdyke Miniature'   Growth dwarf, columnar.
'Westermannii'   Growth conical, dense; leaves yellow to light green.
'White Spot'   Growth columnar; leaves partly white.
'Winston Churchill'   Growth conical; leaves with yellow patches.
'Wisselii'   Growth conical, slender; twigs thick, crowded.
'Wisselii Nana'   Growth dwarf, conical.
'Yellow Transparent'   Leaves yellow in summer, brownish in winter.
'Youngii'   Growth conical; branches stout; leaves glossy green.

Chamaecyparis lawsoniana Cultivar Character Grouping

Dwarf

'Aurea Densa'	'Forsteckensis'	'Nana Albospica'
'Chilworth Silver'	'Gimbornii'	'Nana Argentea'
'Compacta'	'Globosa'	'Nana Glauca'
'Croftway'	'Gnome'	'Nana Rogersii'
'Dwarf Blue'	'Grandis'	'Nidiformis'
'Ellwood's Pygmy'	'Knowefeldensis'	'Pygmaea'
'Ellwood's White'	'Little Spire'	'Pygmaea Argentea'
'Erecta Aurea'	'Lutea Nana'	'Rogersii'
'Ericoides'	'Minima'	'Silver Thread'
'Filiformis Compacta'	'Minima Aurea'	'Tharandtensis'
'Fleckalwood'	'Minima Glauca'	'Tharandtensis Caesia'
'Fletcheri Nana'	'Monumentalis Glauca'	'Wansdyke Miniature'
'Fletcher's Compact'	'Nana'	'Wisselii Nana'

2.   *C. nootkatensis* (D. Don) Spach.   Alaska-cedar.   Nootka-cypress.   Yellow-cedar.   Yellow-cypress.

A false-cypress from northwestern United States. Handsome tree in cultivation but may take several years to establish itself. Lateral leaves about the same size as the corresponding facial leaf. White markings lacking and sprays flattened as in some other species of the genus. Cone 12 mm (½ in) in diameter. Cone scales 4–6.

'Alba'
'Aurea'   Growth conical; leaves yellow to yellow-green.
'Aureovariegata'   Leaves with patches of yellow.

'Compacta'   Growth dwarf, rounded, dense.

'Compacta Glauca'   Similar to 'Compacta'.

'Glauca'   Growth conical, wide, rapid; leaves glaucous.

'Glauca Aureovariegata'   Growth conical; leaves glaucous, variegated yellow.

'Glauca Compacta'   Foliage bluish.

'Glauca Vera'   Growth conical; leaves small, glaucous.

'Lutea'   Young leaves yellow.

'Nidifera'   Growth dwarf, compact.

'Nutans'   Growth conical; similar to 'Glauca'.

'Pendula'   Branches drooping.

'Pendula Variegata'   Branches drooping; leaves with white patches.

'Pendula Vera'   Branches drooping.

'Tatra'   Growth conical; branches many; similar to 'Glauca'.

'Variegata'   Leaves bluish with white patches.

'Viridis'   Growth conical; leaves bright green.

3.   *C. obtusa* (Sieb. & Zucc.) Endl.   Hinoki-cypress.

A large tree, native to Japan. Leaves blunt, the lateral leaves longer than the facial leaf, with some white markings beneath the glands. Cones 13 mm (½ in) in diameter. Cone scales 8; 3–5 narrow-winged seeds per scale.

'Albospica'   Growth slow, compact; young shoots white in spring.

'Aonokujahuhiba'   Similar to 'Filicoides'.

'Argentea'   Leaves white.

'Aurea'   Growth conical; leaves yellow.

'Aurea Nana'   Growth dwarf; foliage yellow.

'Barkenny'   Growth dwarf.

'Bassett'   Similar to 'Juniperoides'.

'Bess'   Growth dwarf, conical.

'Buttonball'   Growth dwarf, rounded.

'Caespitosa'   Growth dwarf, cushionlike.

'Chabo-yadori'   Growth dwarf; some leaves needlelike.

'Chilworth'   Growth dwarf; twigs curving down.

'Chimohiba'   Similar to 'Pygmaea'.

'Compact Fernspray'   Similar to 'Filicoides'.

'Compact Pyramid'   Growth conical, dense; spray tips brown.

'Compacta'   Growth dwarf, conical, dense.

'Compacta Nana'   Same as 'Nana Compacta'.

'Compressa'

'Contorta'   Growth conical; branches twisted.

'Coralliformis'   Growth dwarf; branches many, thick and twisted; tips flattened.

'Coralliformis Nana'   Similar to 'Coralliformis'.

'Crippsii'   Growth slow, conical, dense, spreading when young; foliage yellow.

'Dainty Doll'

'Densa'   Growth dwarf; branches crowded.

'Draht'   Growth conical, compact; leaves crowded.

'Elf'

'Ellie B'

'Erecta'   Growth conical; branches erect; twigs short.

'Ericoides'   Growth dwarf, rounded, dense; foliage gray to light green.

'Fernspray Gold'   Growth dwarf; branches twisted; foliage yellow, similar to 'Kojolkohiba'.

'Filicoides'   Shrub; leaves small, glossy green, bronze in winter.

'Filicoides Compacta'

'Filicoides Graciosa'

'Flabelliformis'   Growth dwarf, rounded.

'Fontana'   Growth conical, wide.

'Gnome'

'Gold Drop'   Growth rounded; foliage yellow.

'Golden Christmas Tree'   Growth slow, rounded; foliage yellow.

'Golden Fairy'   Growth slow, rounded; foliage yellow.

'Golden Nymph'   Growth slow, rounded; foliage yellow.

'Golden Sprite'   Growth slow, rounded; foliage yellow.

'Gold Filament'

'Goldilocks'   Growth conical, dense.

'Gold Spire'   Growth conical, dense; spray tips yellow.

'Gracilis'   Growth conical, compact; leaves short.

'Gracilis Aurea'   Growth conical; leaves yellow to green.

'Gracilis Compacta'   Growth slow, wider.

'Gracilis Nana'   Growth dwarf, irregular.

'Graciosa'   Growth dwarf, compact, robust; leaf tips turned down, creating a "fringed" appearance; similar to 'Nana Gracilis'.

'Grandi Pygmaea'

'Green Cushion'

'Green Diamond'   Growth dense; twigs fused to form fan-shaped foliage and twigs.

'Hage'   Growth dwarf, conical, dense, compact.

'Hoseri'

'Intermedia'   Growth dwarf, conical.

'Junior'   Growth slow, dense, spherical.

'Juniperoides'   Growth dwarf, rounded.

'Juniperoides Compacta'   Growth dwarf, rounded, dense.

'Kaanamihiba'   Growth dwarf, wider than high; branches thick; leaves yellow, curved down.

'Kamaeni Hiba'   Growth dwarf; twigs threadlike; foliage yellow.

'Kamakura Hiba'   Growth dwarf; branches short, twigs and leaves crowded.

'Kojolkohiba'   Growth dwarf, conical; foliage yellow, crowded.

'Kosteri'   Growth dwarf, conical; branches horizontal or erect.

'Kosteri Nana'   Growth dwarf, irregular; foliage lacy.

'Laxa'   Similar to 'Nana Gracilis'.

'Leprechaun'

'Little Marky'

'Loughead'

'Lutea Nova'   Growth conical; twig tips yellow, turned down.

'Lycopodioides'   Growth loose, wide, and random; some leaves awl-shaped, blue-green.

'Lycopodioides Aurea'   Growth slow; branches slender; some leaves awl-shaped, yellow.

'Magnifica'   Growth conical.

'Mariesii'   Growth slow, conical, dense to open; foliage variegated white or yellow.

'Mimima'   Growth dwarf, cushionlike, dense; branches and twigs four-sided, similar to 'Caespitosa'.

'Nana'   Growth dwarf, rounded; leaves small.

'Nana Argentea'   Growth dwarf, rounded; leaves variegated white.

'Nana Aurea'   Growth dwarf, conical; foliage variegated yellow and white.

'Nana Compacta'   Growth dwarf, rounded, dense.

'Nana Contorta'   Growth dwarf, rounded, dense; branches twisted.

'Nana Densa'   Similar to 'Nana'.

'Nana Gracilis'   Growth dwarf; branches often twisted.

'Nana Kosteri'   Growth dwarf, compact.

'Nana Lutea'   Growth dwarf, spreading when young but higher than wide; leaves yellow.

'Nana Prostrata'   Growth dwarf, spreading over the ground.

'Nana Pyramidalis'   Growth dwarf, conical, dense.

'Nana Repens'

'Opaal'   Growth dwarf; leaves yellow-green.

'Prostrata'   Growth slow, low, flat-topped.

'Pygmaea'   Growth dwarf, wide; foliage reddish.

'Pygmaea Aurescens'   Growth dwarf, wide; leaves bronze on upper surface.

'Pygmaea Densa'   Similar to 'Pygmaea'.

'Rainbow'

'Reis'   Same as 'Reis Dwarf'.

'Reis Dwarf'   Growth dwarf, conical; long shoots often produced; leaves in tufts.

'Repens'   Growth dwarf, spreading over the ground.

'Rezek'

'Rigid Dwarf'   Growth dwarf, dense, stiff, twice as high as wide.

'Rigida'

'Sanderi'   Growth slow; leaves awl-shaped, in threes and twos.

'Snowkist'   Growth dwarf; leaves variegated white; similar to 'Tonia'.

'Spiralis'   Growth dwarf; twigs twisted.

'Split Rock'   Growth dwarf; leaves awl-shaped.

'Stoneham'   Growth dwarf; similar to 'Nana'.

'Suirova-hiba'   Similar to 'Coralliformis'.

'Sunburst'

'Templehof'   Shrub; growth compact, conical.

'Tetragona'   Growth dwarf, conical, compact; twigs crowded.

'Tetragona Aurea'   Growth dwarf, conical, compact; twigs crowded; leaves yellow.

'Tetragona Intermedia'

'Tetragona Minima'   Growth dwarf, cushionlike.

'Tiny Tot'   Growth dwarf, rounded.

'Tonia'   Growth dwarf; leaves variegated white.

'Torulosa'   Growth spreading when young, becoming conical; twigs twisted, threadlike.

'Torulosa Nana'   Growth dwarf, rounded; twigs twisted, threadlike.

'Tsatsumi'   Growth dwarf; twigs somewhat threadlike; similar to 'Coralliformis'.

'Van Ness'

'Verbanensis'

'Verdoni'   Growth dwarf; foliage yellow, becoming bronze.

'Wells Special'   Growth slow; leaves dark green.
'White Tip'   Similar to 'Nana Argentea'.
'Yellowtip'   Growth dwarf; young leaves yellow.
'Youngii'   Twigs drooping; leaves yellow.

Chamaecyparis obtusa Cultivar Character Groupings

Dwarf

'Aurea Nana'	'Kaanamihiba'	'Pygmaea'
'Barkenny'	'Kamaeni Hiba'	'Reis'
'Bassett'	'Kojolkohiba'	'Reis Dwarf'
'Bess'	'Kosteri'	'Repens'
'Buttonball'	'Kosteri Nana'	'Rigid Dwarf'
'Caespitosa'	'Laxa'	'Snowkist'
'Chabo-yadori'	'Minima'	'Spiralis'
'Chilworth'	'Nana'	'Split Rock'
'Chimohiba'	'Nana Argentea'	'Stoneham'
'Compacta'	'Nana Aurea'	'Suirova-hiba'
'Coralliformis'	'Nana Compacta'	'Tetragona'
'Coralliformis Nana'	'Nana Contorta'	'Tetragona Aurea'
'Densa'	'Nana Densa'	'Tetragona Minima'
'Ericoides'	'Nana Gracilis'	'Tiny Tot'
'Fernspray Gold'	'Nana Kosteri'	'Tonia'
'Flabelliformis'	'Nana Lutea'	'Torulosa Nana'
'Gracilis Nana'	'Nana Prostrata'	'Tsatsumi'
'Graciosa'	'Nana Pyramidalis'	'Verdoni'
'Hage'	'Nana Repens'	'White Tip'
'Intermedia'	'Opaal'	'Yellowtip'
'Juniperoides'	'Pygmaea Aurescens'	
'Juniperoides Compacta'	'Pygmaea Densa'	

Yellow

'Aurea'	'Golden Sprite'	'Nana Aurea'
'Aurea Nana'	'Gold Spire'	'Nana Lutea'
'Crippsii'	'Gracilis Aurea'	'Opaal'
'Fernspray Gold'	'Kamakura Hiba'	'Tetragona Aurea'
'Gold Drop'	'Kojolkohiba'	'Verdoni'
'Golden Christmas Tree'	'Lutea Nova'	'Yellowtip'
'Golden Fairy'	'Lycopodiodes Aurea'	'Youngii'
'Golden Nymph'	'Mariesii'	

White

| 'Albospica' | 'Nana Argentea' | 'Tonia' |
| 'Argentea' | 'Snowkist' | 'White Tip' |

4. *C. pisifera* **(Sieb. & Zucc.) Endl.** **Sawara-cypress.**

A large tree native to Japan, with distinctive, strongly flattened sprays spaced at even intervals in horizontal planes. Leaves sharp-pointed, often conspicuously so, with prominent white markings beneath. The leaves of the 'Squarrosa' and 'Plumosa' groups and several of the other cultivars all awl-shaped. Cones 7–10 mm ($\frac{1}{4}$–$\frac{3}{8}$ in) in diameter. Scales 10, sometimes 12, with 1–2 broad, winged seeds per scale.

'Argentea'
'Argentea Nana'   Growth rounded, dense; foliage soft, feathery.
'Argenteovariegata'   Growth slow; leaves variegated yellow.
'Aurea'   Growth conical; leaves yellow at ends of the twigs.
'Aurea Compacta Nana'   Growth slow, mounded; leaves yellow at twig ends.
'Aurea Nana'   Growth dwarf, conical; leaves yellow.
'Aureovariegata'   Growth slow; leaves variegated yellow.
'Boulevard'   Growth slow, conical; leaves awl-shaped.
'Clouded Sky'   Growth conical; leaves awl-shaped.
'Compacta'   Growth dwarf, rounded, twice as wide as high.
'Compacta Albovariegata'   Similar to 'Compacta' but leaves variegated white or yellow.
'Compacta Nana'   Growth slower and more dense than 'Compacta'.
'Compacta Variegata'   Growth dwarf, rounded; leaves variegated white.
'Compressa Variegata'
'Cream Ball'   Growth slow; leaves awl-shaped, white.
'Cyanoviridis'   Same as 'Boulevard'.
'Dwarf Blue'   Similar to 'Squarrosa Intermedia'.
'Ericoides'   Growth dwarf, conical, dense.
'Filifera'   Growth conical or rounded; sprays threadlike, drooping; leaves awl-shaped.
'Filifera Argenteovariegata'   Growth dwarf; leaves variegated white.

'Filifera Aurea' Growth conical, sprays threadlike; leaves awl-shaped, yellow.

'Filifera Aurea Nana' Growth dwarf; sprays threadlike; leaves awl-shaped, yellow; may be the same as 'Golden Mop'.

'Filifera Aureovariegata' Growth dwarf, conical; branches threadlike; foliage variegated yellow.

'Filifera Gracilis' Similar to 'Filifera'.

'Filifera Nana' Growth dwarf; leaves awl-shaped.

'Filifera Pendula'

'Filifera Variegata' Similar to 'Filifera' but variegated white.

'Glauca Compacta Nana' Growth dwarf, dense, rounded; branches drooping at tips; foliage blue.

'Globosa' Growth rounded.

'Gold Dust' Growth dwarf, rounded to conical; foliage tipped yellow.

'Golden Mop' Growth dwarf, rounded when young, conical with age; branches drooping at tips; leaves yellow.

'Gold Spangle' Growth conical, dense; leaves yellow.

'Green Velvet' Similar to 'Squarrosa'.

'Greg's Sport' Growth rapid; leaves awl-shaped, tufted.

'Hime-savara' Growth dwarf, rounded; branches threadlike.

'Juniperoides Aurea' Growth dwarf; twigs crowded; leaves crowded, yellow.

'Leptoclada'

'Lombards' Leaves awl-shaped, bronze or gray.

'Lutea' Growth dwarf; leaves yellow.

'Lutescens' Growth conical; branches ascending; sprays feathery; foliage variegated yellow.

'Mikko' Same as 'Snow'.

'Mikko Snow' Same as 'Snow'.

'Minima' Same as 'Nana'.

'Minima Aurea' Growth dwarf, compact; foliage yellow.

'Minima Variegata' Growth dwarf; foliage variegated yellow.

'Monstrosa'

'Mops' Same as 'Golden Mop'.

'Nana' Growth dwarf and dense.

'Nana Albovariegata' Growth dwarf; foliage occasionally variegated white.

'Nana Aurea' Growth dwarf; yellow.

'Nana Aureovariegata' Growth dwarf, rounded; leaves variegated yellow.

'Nana Compacta'   Growth dwarf.

'Nana Variegata'   Similar to 'Compacta Variegata'.

'Parslori'   Similar to 'Nana'.

'Pendula'   Growth rapid; branches drooping.

'Pici'   Similar to 'Squarrosa Dumosa'.

'Plumosa'   Growth conical; leaves awl-shaped.

'Plumosa Albopicta'   Growth conical; twigs feathery; leaves awl-shaped, white at twig tips.

'Plumosa Argentea'   Growth conical; leaves awl-shaped, sometimes silver.

'Plumosa Aurea'   Growth conical; leaves awl-shaped, yellow.

'Plumosa Aurea Compacta'   Growth conical; leaves awl-shaped, yellow, larger than 'Plumosa'.

'Plumosa Aurea Nana'   Growth dwarf; leaves awl-shaped.

'Plumosa Compacta'   Growth conical, compact; twigs crowded; some leaves awl-shaped.

'Plumosa Compressa'   Growth dwarf, wider than high; twigs crowded; leaves crowded, awl-shaped.

'Plumosa Compressa Aurea'   Growth dwarf; leaves awl-shaped, yellow.

'Plumosa Cream Ball'   Growth slow; leaves awl-shaped, white.

'Plumosa Cristata'   Growth dwarf, compact; leaves awl-shaped.

'Plumosa Flavescens'   Growth dwarf; rounded or conical; leaves awl-shaped, sometimes yellow.

'Plumosa Juniperoides'   Similar to 'Plumosa Compressa' but leaves pointing forward along the twig.

'Plumosa Lutescens'   Leaves awl-shaped, yellow.

'Plumosa Minima Variegata'

'Plumosa Nana Aurea'   Growth dwarf, dense; leaves awl-shaped, yellow.

'Plumosa Nana Variegata'   Growth dwarf, rounded; leaves awl-shaped.

'Plumosa Pygmaea'   Similar to 'Plumosa Rogersii' but foliage green.

'Plumosa Rogersii'   Growth dwarf, higher than wide; branches erect; leaves awl-shaped, yellow.

'Plumosa Vera'   Growth conical, dense; leaves awl-shaped.

'Pygmy'

'Silver Lode'   Growth dwarf, rounded, twice as wide as high; foliage white.

'Snow'   Growth dwarf, rounded; leaves awl-shaped, white at twig tips.

'Sopron'   Growth low, wide, spreading; leaves awl-shaped.

'Squarrosa'   Growth conical; leaves awl-shaped.

'Squarrosa Argentea'   Growth slow; leaves awl-shaped.

'Squarrosa Argentea Compacta'

'Squarrosa Aurea'   Growth slow, conical; sprays feathery; leaves awl-shaped.

'Squarrosa Aurea Nana'   Similar to 'Squarrosa Lutea'.

'Squarrosa Cristata'   Growth dwarf, compact; leaves awl-shaped, blue.

'Squarrosa Cyanoviridis'   Growth slow, compact; foliage silvery.

'Squarrosa Dumosa'   Growth conical or rounded; leaves awl-shaped, bronze in winter.

'Squarrosa Elegans'   Growth compact; leaves awl-shaped, yellow.

'Squarrosa Intermedia'   Growth dwarf; leaves awl-shaped, in threes, glaucous.

'Squarrosa Juniperoides'

'Squarrosa Lutea'   Growth conical; leaves awl-shaped, yellow to white.

'Squarrosa Minima'   Growth slow, rounded; leaves awl-shaped, in twos and threes, small, pale.

'Squarrosa Monstrosa'

'Squarrosa Monstrosa Nana'   Similar to 'Greg's Sport'.

'Squarrosa Nana'

'Squarrosa Pygmaea'   Growth dwarf, rounded becoming conical, dense; leaves awl-shaped, blue, similar to 'Minima'.

'Squarrosa Sieboldii'   Growth dwarf, rounded; twigs feathery; leaves awl-shaped.

'Squarrosa Sulphurea'   Growth conical; twigs feathery; leaves awl-shaped, yellow.

'Strathmore'   Growth dwarf, rounded, spreading; leaves yellow.

'Sulphurea'   Leaves yellow.

'Sulphurea Nana'   Growth dwarf, rounded; foliage yellow.

'Sungold'   Similar to 'Filifera Aurea'.

'Tamarawardiana'

'Tsukumi'   Similar to 'Nana'.

'Westermannii'

'White Pygmy'   Similar to 'Nana' but leaves white at twig tips.

'Winter Gold'   Leaves variegated yellow; similar to 'Compacta Variegata'.

Chamaecyparis pisifera Cultivar Character Groupings

Dwarf

'Aurea Nana'	'Compacta Nana'	'Dwarf Blue'
'Compacta'	'Compacta Variegata'	'Ericoides'

'Filifera Argenteovariegata'
'Filifera Aurea Nana'
'Filifera Aureovariegata'
'Filifera Nana'
'Glauca Compacta Nana'
'Gold Dust'
'Golden Mop'
'Hime-savara'
'Juniperoides Aurea'
'Lutea'
'Mikko'
'Minima'
'Minima Aurea'
'Minima Variegata'
'Mops'

'Nana'
'Nana Albovariegata'
'Nana Aurea'
'Nana Aureovariegata'
'Nana Compacta'
'Nana Variegata'
'Parslori'
'Plumosa Aurea Nana'
'Plumosa Compressa'
'Plumosa Compressa Aurea'
'Plumosa Cristata'
'Plumosa Flavescens'
'Plumosa Juniperoides'
'Plumosa Nana Aurea'
'Plumosa Nana Variegata'

'Plumosa Pygmaea'
'Plumosa Rogersii'
'Pygmy'
'Silver Lode'
'Snow'
'Squarrosa Cristata'
'Squarrosa Intermedia'
'Squarrosa Nana'
'Squarrosa Pygmaea'
'Squarrosa Sieboldii'
'Strathmore'
'Sulphurea Nana'
'Tsukumi'
'White Pygmy'
'Winter Gold'

Yellow

'Argenteovariegata'
'Aurea'
'Aurea Compacta Nana'
'Aurea Nana'
'Aureovariegata'
'Compacta Albovariegata'
'Filifera Aurea'
'Filifera Aurea Nana'
'Filifera Aureovariegata'
'Gold Dust'
'Golden Mop'
'Gold Spangled'
'Juniperoides Aurea'
'Lutea'

'Lutescens'
'Mikko'
'Minima Aurea'
'Minima Variegata'
'Mops'
'Nana Aurea'
'Nana Aureovariegata'
'Plumosa Aurea'
'Plumosa Aurea Compacta'
'Plumosa Compressa Aurea'
'Plumosa Flavescens'
'Plumosa Lutescens'
'Plumosa Nana Aurea'

'Plumosa Rogersii'
'Snow'
'Squarrosa Aurea'
'Squarrosa Aurea Nana'
'Squarrosa Elegans'
'Squarrosa Lutea'
'Squarrosa Sulphurea'
'Strathmore'
'Sulphurea'
'Sulphurea Nana'
'Sungold'
'Winter Gold'

White

'Compacta Variegata'
'Cream Ball'
'Filifera Argenteovariegata'

'Filifera Variegata'
'Nana Albovariegata'
'Nana Variegata'
'Plumosa Albopicta'

'Silver Lode'
'Snow'
'White Pygmy'

+**5.   *C. thyoides* (L.) BSP.   White-cedar.**

Native tree in the Northeast that does not make as good an ornamental as other species of the genus. Does not perform well in windy or dry locations. Cones 7 mm (¼ in) in diameter. Cone scales 6.

'Andelyensis'   Growth slow, conical, compact; branches short, thick; twigs crowded; some young leaves awl-shaped.

'Andelyensis Aurea'   Growth dwarf, wider than high; otherwise similar to 'Andelyensis'.

'Andelyensis Conica'   Growth dwarf; twigs crowded, leaves awl-shaped, in threes.

'Andelyensis Nana'   Growth dwarf, rounded, then spreading.

'Atrovirens'   Leaves dark green.

'Aurea'   Leaves yellow.

'Conica'   Growth dwarf, conical, dense; leaves awl-shaped.

'Ericoides'   Growth dwarf; leaves awl-shaped, soft to touch.

'Glauca'   Leaves tinted white.

'Golden Twig'

'Heatherbun'   Growth slow; leaves small, purple in winter.

'Hopkinton'   Growth columnar, rapid; cones numerous.

'Hoveyi'   Twigs crowded together into tufts.

'Little Jamie'   Growth dwarf, columnar; foliage soft, partially purplish in winter.

'Nana'   Growth dwarf, wider than high; leaves blue-green.

'Pygmaea'   Growth dwarf, cushionlike, spreading.

'Rezek's Dwarf'   Same as 'Little Jamie'.

'Rubicon'   Growth dwarf, conical.

'Variegata'   Leaves variegated yellow.

Cryptomeria japonica

Cryptomeria D. Don.   Cryptomeria.
Japanese-cedar.

This genus contains only one species. It is native to eastern Asia but is hardy in northeastern North America.

1. *C. japonica* D. Don.   Cryptomeria.   Japanese-cedar.

Native to Japan and China. An unusual tree planted for ornament, often close to buildings. Red to reddish-brown bark that shreds or peels off in strips. Leaves angled, keeled, awl-shaped, 0.7–2.5 cm (¼–1 in) long, green usually turning bronze in winter. Cones with spiny appearance, 2.0–2.5 cm (¾–1 in) in diameter.   var. *sinensis* Sieb. & Zucc. Native to China; branches thin; leaves long and thin; cones with fewer scales.

'Albospica'   Leaves at twig tips on young plants white.
'Albovariegata'   Growth dwarf; leaves at branch tips white.

'Araucarioides'   Growth dwarf, conical, or irregular; branches long, stiff, whiplike; twigs drooping; leaves short.

'Aurea'   Leaves yellow.

'Aureovariegata'   Some leaves variegated yellow.

'Aurescens'   Growth conical; branches crowded; leaves yellow-green.

'Bandai-sugi'   Growth slow, rounded; twigs and leaves longer and shorter.

'Birodo-sugi'   Similar to 'Compressa'.

'Chabo-sugi'   Same as 'Nana'.

'Compacta'   Growth slow, conical, compact; branches feathery at tips.

'Compacta Nana'   Growth dwarf, conical.

'Compressa'   Growth conical to rounded; branches short, crowded; twig tips curled; leaves crowded.

'Cristata'   Growth conical; twigs short, stiff, many fused into fan-shaped growths; leaves may be yellow or tipped white.

'Dacrydioides'   Shrub, with many stems, twigs in groups at the ends of the branches.

'Eizan-sugi'   Growth dwarf.

'Elegans'   Branches horizontal; leaves brownish red in winter.

'Elegans Aurea'   Branches drooping; foliage yellow-green in winter.

'Elegans Compacta'   Growth dwarf, compact; branches horizontal.

'Elegans Nana'   Growth dwarf; forming broad mounds; foliage purplish blue in winter.

'Elegans Viridis'   Leaves remaining green in winter.

'Enko-sugi'   Same as 'Araucarioides'.

'Fasciata'   Growth dwarf, low; twigs few.

'Filifera'   Branches long.

'Giokumo'   Shrub; growth slow.

'Globosa'   Growth rounded, wider than high; branches crowded; leaves thick.

'Globosa Nana'   Growth dwarf, rounded; leaves numerous, short.

'Gracilis'   Growth slender; leaves short.

'Gyokruya'   Growth slow, irregularly rounded.

'Hoo-sugi'   Same as 'Selaginoides'.

'Ito-sugi'   Branches scarce, thin, crowded near top of tree; twigs crowded at tips of branches; similar to 'Dacrydioides'.

'Jindai-sugi'   Growth conical; leaves short.

'Kewensis'   Growth dwarf; twigs short, at right angles to the branches.

'Kilmacurragh'   Growth dwarf; twigs fused into fan-shaped growths.

'Knaptonensis'   Growth dwarf, compact; leaves short, pale, does not perform well in frost and strong sun.

'Kusari-sugi'   Same as 'Spiralis'.

'Lobbii'   Growth conical, compact; common.

'Lobbii Compacta'

'Lobbii Nana'   Growth slow, flat-topped; branches thick; similar to 'Elegans Nana'.

'Lycopodioides'   Shrub; growth open, loose; branches long, snakelike; leaves short.

'Mankichi-sugi'   Similar to 'Monstrosa Nana'.

'Mankitiana-sugi'   Similar to 'Monstrosa'.

'Mejero-sugi'   Growth dwarf, compact.

'Monstrosa'   Growth columnar, with many short branches near top; leaves on older branches very long.

'Monstrosa Nana'   Growth dwarf, columnar.

'Nana'   Growth dwarf; branches erect and twisted, the tips drooping; twigs crowded and stiff.

'Nana Albospica'   Growth dwarf; young leaves white.

'Ogon-sugi'   Same as 'Aurea'.

'Okina-sugi'   Similar to 'Nana Albospica'.

'Osaka-tama-sugi'   Similar to 'Vilmoriniana'.

'Pendulata'   Branches long, slender, drooping.

'Pungens'   Growth dwarf, compact; leaves stiff, sharp-pointed, short, dark green.

'Pygmaea'   Growth dwarf; twigs drooping.

'Rein's Dense Jade'   Growth dwarf, rounded; foliage green.

'Sekkan-sugi'   Same as 'Cristata'.

'Sekko-sugi'   Same as 'Cristata'.

'Sekkwa-sugi'   Same as 'Cristata'.

'Selaginoides'   Branches long, slender, whiplike, with twigs in clusters at the ends.

'Shishi-gashira'   Similar to 'Monstrosa Nana'.

'Spiralis'   Growth rounded, compact; leaves twisted around twig.

'Spiraliter Falcata'   Growth taller than 'Spiralis'; branches longer, thinner, twisted.

'Tansu'   Growth dwarf, congested.

'Variegata'   Foliage with patches of yellow.

'Vilmoriniana'   Growth dwarf, dense, rounded; branches short; twigs crowded; leaves reddish purple in winter.

'Viminalis'   Branches slender, with tufted twigs near the tips.

'Yatsubusa'   Same as 'Tansu'.

'Yatsubusa-sugi'   Similar to 'Elegans Nana'.

'Yatsufusa'   Same as 'Tansu'.

'Yawara-sugi'   Similar to 'Elegans'.

'Yore-sugi'   Growth rounded, compact; leaves short; similar to 'Spiralis'.

'Yoshino'   Growth columnar, rapid.

Cryptomeria japonica Cultivar Character Grouping

Dwarf

'Albovariegata'	'Kilmacurragh'	'Okina-sugi'
'Chabo-sugi'	'Knaptonensis'	'Osaka-tama-sugi'
'Compacta Nana'	'Lobbii Nana'	'Pygmaea'
'Eizan-sugi'	'Mankichi-sugi'	'Rein's Dense Jade'
'Elegans Compacta'	'Mankitiana-sugi'	'Shishi-gashirad'
'Elegans Nana'	'Mejero-sugi'	'Tansu'
'Enko-sugi'	'Monstrosa Nana'	'Vilmoriniana'
'Globosa Nana'	'Nana'	'Yatsubusa'
'Kewensis'	'Nana Albospica'	'Yatsubusa-sugi'

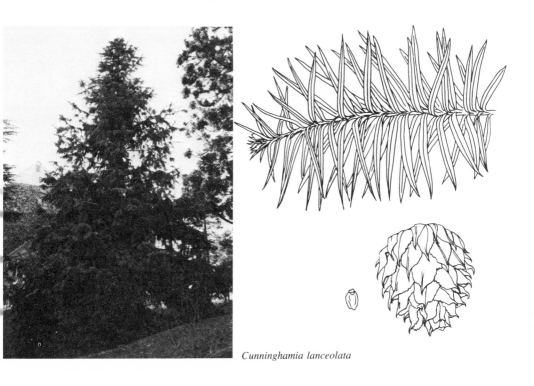

Cunninghamia lanceolata

Cunninghamia R. Br.   China-fir.

This genus contains three species native to eastern Asia, but only one is hardy in the Northeast.

1.   *C. lanceolata* (Lamb.) Hook.   China-fir.

Tree, native to China, barely hardy in the Northeast, thus cultivated specimens often displaying many brown leaves. Leaves quite broad, particularly at the base, blue-gray, 3.5–6.5 cm (1¼–2½ in) long. Cones 2.5–5.0 cm (1–2 in) long.

'Compacta'   Growth dwarf without leading shoot.
'Glauca'   Leaves blue-green.

×Cupressocyparis leylandii

× Cupressocyparis Dallim. (*Cupressus* L. × *Chamaecyparis* Spach.) Leyland-cypress.

This genus is a hybrid with one species.

1. *×C. leylandii* (Jacks. & Dallim.) Dallim. (*Cupressus macrocarpa Hartweg* × *Chamaecyparis nootkatensis* (D. Don) Spach.) Leyland-cypress.

Chance hybrid from cultivated plants of *Chamaecyparis nootkatensis* and *Cupressus macrocarpa*. Obvious characters mostly those of *Chamaecyparis nootkatensis,* including hardiness in the Northeast. Cones 2 cm (¾ in) in diameter, 5 seeds for each of the 8 cone scales. Difficult to distinguish this species from *Chamaecyparis* even with cones. An excellent ornamental.

'Castlewellan Gold' Leaves yellow.
'Green Spire' Growth columnar, compact.

74

'Haggerston Grey' Growth more open; twigs regularly spaced.
'Leighton Green' Growth columnar; twigs irregularly spaced.
'Naylor's Blue' Leaves grayish blue; spray tips flat.
'Robinson's Gold' Branches crowded; leaves yellow.
'Silver Dust' Growth columnar; spray tips white.
'Stapehill' Growth columnar.

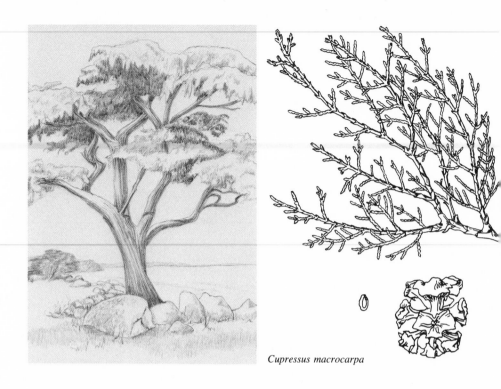

Cupressus macrocarpa

Cupressus L. Cypress.

This consists of about 15 species in Asia, the Mediterranean, and America, none of which is native to northeastern North America and only two of which can be seen in cultivation and then only rarely. Cypresses are often shrubby plants (unlike the California species illustrated) with small scalelike leaves, the edges of which are minutely toothed (viewed with a magnifying glass or hand lens). This character distinguishes *Cupressus* from *Thuja* and *Chamaecyparis*. *Cupressus* has spherical cones with shield-shaped scales, as does *Chamaecyparis*, but most of the 6–14 scales bear many seeds per scale, and none are borne on the lower scales.

1. Leaves dark green, the tips tending to be blunt . . . . . *C. macnabiana*
   Macnab cypress
1. Leaves gray, the tips sharp-pointed . . . . *C. arizonica*
   Arizona cypress

1. *C. arizonica* Greene.   Arizona cypress.

Small tree native to southwestern United States and Mexico. Cones 2.0–2.5 cm (¾–1 in) in diameter.

'Compacta' Growth dwarf, rounded to conical, dense.

'Conica' Growth conical; twigs numerous, short, stiff; leaves crowded.

'Crowborough' Similar to 'Compacta'.

'Fastigiata' Growth columnar.

'Fastigiata Aurea' Growth columnar; foliage with some yellow coloring.

'Gareei' Growth dwarf, columnar; sunny, windless, well-drained environment probably necessary.

2. *C. macnabiana* A. Murr. Macnab cypress. McNab cypress.

Shrub native to the canyons, hillsides, and coastal mountains of California. Leaves thickened at the blunt tips. Cones 2.0–2.5 cm (¾–1 in) in diameter.

'Sulphurea' Leaves at twig tips yellow.

Juniperus rigida

Juniperus virginiana

Juniperus virginiana

Juniperus L.   Juniper.

This genus spans the Northern Hemisphere and encompasses about 60 species of trees and shrubs and low, spreading plants with a multitude of varieties that serve many horticultural purposes. The junipers are most popular as ground covers (particularly on slopes) and as wall climbers or for plantings close to buildings. These plants differ from most other conifers in that they have a reproductive structure that is about 10 mm (⅓ in) in diameter and resembles a berry more than a cone. The leaves, often a bluish or light green, are awl-shaped and needlelike or scalelike. The young shoots bear needlelike leaves, which are usually about 1–2 cm (⅓–¾ in) long and have white markings on the upper surface, a character that distinguishes the genus from *Cupressus*. Many of the species are extremely difficult for observers to identify who lack familiarity with the genus and do not observe the plants side by side.

79. *J. rigida*

1. Leaves all needlelike.
    2. Leaves joining the twig in one place, with solid white longitudinal band on upper surface (often hard to see because of the concave or furrowed nature of the leaves) (Fig. 79).

78

3. Leaves deeply grooved on upper surface so that the white longitudinal band is hidden; white band usually narrower than either of the green edges.

    4. Plants spreading and low . . . . *J. conferta*
                                  Shore juniper

    4. Plants erect; trees or shrubs . . . . *J. rigida*
                                    Needle juniper

3. Leaves merely concave on upper surface, the white longitudinal band as wide or wider than the green edges . . . . *J. communis* *+
                                    Common juniper

2. Leaves joining the twig for some distance along stem (Fig. 80).

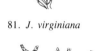

80. *J. squamata*

    5. Leaves with 2 white spots at the base, glaucous . . . . *J. procumbens*
                                   Creeping juniper

    5. Leaves lacking white spots at the base, green or blue . . . . *J. squamata*
                                    Himalayan juniper

1. Leaves mostly scalelike.

  6. Twigs, when crushed, emitting strong odor; berrylike structures hanging on curved stalks.

    7. Leaves dark green with disagreeable odor . . . . *J. sabina*
                                       Savin

    7. Leaves blue-green with agreeable odor . . . . . . *J. horizontalis* *+
                                      Creeping juniper

  6. Twigs lacking strong odor.

    8. Needlelike leaves opposite (may be in whorls of 3 on leading shoots); tips of scalelike leaves may be sharp (Fig. 81); berrylike structure blue.

81. *J. virginiana*

      9. Sprays spreading; bark shredding . . . . *J. virginiana* *+
                                 Eastern red-cedar

      9. Sprays shorter, closer together; bark loose but not falling . . . . *J. scopulorum*
                               Rocky Mountain juniper

    8. Needlelike leaves sometimes in whorls of 3 on small clusters of twigs; tips of scalelike leaves tending to be blunt

82. *J. chinensis*

(Fig. 82); berrylike structure brown . . . . *J. chinensis**
                                  Chinese juniper

*1. *J. chinensis* L.   Chinese juniper.

Native to southeast Asia, commonly planted, with many forms. Can grow to be a tree but in northeastern North America most often found as a shrub. Most leaves on the mature branches scalelike, obtuse, and lacking strong odor when crushed. Needlelike leaves present in small clusters on the twigs and mostly in threes. Male cones numerous, yellow, and often found on young shoots.

var. *Sargentii* A. Henry. Sargent juniper. Shrub; branches resting on ground with twigs ascending, forming dense mats; some leaves awl-shaped or needlelike. Five cultivars of this botanical variety: 'Compacta', growth compact; 'Glauca', growth dwarf; twigs thin and feathery; 'Manay', foliage blue-green; 'Variegata', leaf tips white; 'Viridis', leaves light green.

'Alba'   Twig tips white.

'Albospica'   Some twig tips yellow, some white.

'Ames'   Growth slow; leaves needlelike.

'Aquarius'   Growth compact; foliage blue-green.

'Arbuscula'   Growth conical.

'Arctic'   Same as 'Pfitzer Arctic'.

'Armstrongii'   Dense shrub.

'Aurea'   Growth conical; branches erect; leaves yellow, some needlelike.

'Aurea Globosa'   Growth rounded; leaves yellow.

'Berry Hill'   Shrub; leaves both needlelike and scalelike.

'Blaauw'   Growth slow, vase-shaped; sprays numerous, feathery; some leaves needlelike.

'Blue Alps'   Growth upright; branches drooping somewhat.

'Blue Cloud'   Growth low; twigs threadlike.

'Blue Point'   Growth regular with many branches; leaves blue-gray.

'Blue Vase'   Growth spreading, upright.

'Columnaris'   Growth columnar, rapid; leaves needlelike.

'Columnaris Glauca'   Growth columnar; leaves needlelike, gray.

'Columnaris Hetzii'   Same as 'Hetzii Columnaris'.

'Compacta'

'Densa'   Growth conical.

'Densa Erecta'

'Dropmore'   Growth dwarf, dense.

'Echiniformis'   Growth dwarf, flattened; twigs short, crowded; some leaves needlelike.

'Excelsior'   Growth columnar; similar to 'Keteleeri'.

'Expansa'

'Fairview'   Growth conical; most leaves needlelike.

'Foemina'   Growth conical; twigs crowded.

'Fortunei'   Growth columnar; leaves mostly needlelike.

'Fruitlandii'   Growth spreading but thickly branched.

'Glauca'   Growth columnar; foliage gray.

'Globosa'   Growth dwarf, rounded; twigs crowded.

'Globosa Cinerea'   Similar to 'Blaauw'.

'Gold Coast'   Leaves yellow at branch tips.

'Hetzii'   Shrub; some leaves needlelike.

'Hetzii Columnaris'   Growth columnar; some leaves needlelike, green, sharp.

'Hetzii Glauca'   Growth fast, upright but spreading.

'Hills Blue'   Similar to 'Pfitzeriana Glauca'.

'Idylwild'   Growth conical, dense, rapid.

'Iowa'   Growth slow; some leaves needlelike.

'Jacobiana'   Growth columnar; branches short; twigs many, fine.

'Japonica'   Growth dwarf; some leaves needlelike, sharp.

'Japonica Aureovariegata'   Growth low; foliage variegated yellow.

'Japonica Variegata'   Growth dwarf; some leaves needlelike, patched with white.

'Kaizuka'   Growth conical, erect, narrow; twigs in tufts along the branches.

'Kaizuka Variegated'   Same as 'Variegated Kaizuka'.

'Kallays Compact'   Growth dwarf; leaves deep green, many of them needlelike.

'Keteleeri'   Growth conical, narrow, dense; twigs crowded.

'Kohankie's Compact'   Growth dwarf, rounded, dense.

'Kosteriana'   Same as *J. virginiana* 'Kosteri'.

'Kuriwao Gold'   Growth upright, dense; foliage yellow.

'Leeana'   Growth columnar; branches crowded.

'Maney'   Growth vase-shaped; leaves needlelike.

'Mas'   Growth upright, loose; some leaves needlelike.

'Mathot'   Growth dwarf; leaves needlelike, opposite, upper side turned outward; similar to 'Kallays Compact'.

'Meyeri'   Foliage very glaucous.

'Mint Julep'   Branches arching; leaves rich green.

'Mission Spire'   Growth columnar, wide.

'Monarch'   Similar to 'Iowa'.

'Moraine'   Growth dense, some leaves needlelike.

'Mordigan'   Growth compact; leaves yellow.

'Mountbatten'   Growth conical, dense, compact; leaves mostly
   needlelike.

'Nana Aurea'

'Neaboriensis'   Growth conical; some leaves needlelike.

'Nick's Compact'   Growth compact, spreading; branches with drooping
   tips.

'Obelisk'   Growth columnar, dense, slender; leaves needlelike,
   upper side turned outward.

'Oblonga'   Plant with one stem; some leaves needlelike, sharp.

'Old Gold'   Growth compact; leaves yellow, some needlelike.

'Olympia'   Growth columnar, slender; some leaves needlelike.

'Ontario Green'   Growth upright; leaves mostly needlelike.

'Owen'   Growth low, compact, dense; some leaves needlelike.

'Parsonii'   Shrub; twigs cordlike.

'Pendula'   Branches drooping.

'Pfitzer Arctic'   Growth low.

'Pfitzer Blue Gold'   Growth spreading; foliage patched yellow.

'Pfitzer Kallay'

'Pfitzer Matthews Blue'   Similar to 'Pfitzeriana'; foliage blue.

'Pfitzer Mordigan'   Similar to 'Pfitzeriana'; foliage yellow.

'Pfitzer Nels'

'Pfitzer Ozark Compact'

'Pfitzer Sarcoxie'   Growth low, spreading, compact.

'Pfitzeriana'   Growth rapid, dense, spreading; branches wavy; some
   leaves needlelike.

'Pfitzeriana Argentea'

'Pfitzeriana Armstrongii'   Leaves needlelike, light green.

'Pfitzeriana Aurea'   Growth dense, wide; some leaves needlelike;
   foliage tinted yellow in summer.

'Pfitzeriana Compacta'   Growth slow, dense; branches wavy; some
   leaves needlelike.

'Pfitzeriana Glauca'   Growth dense; leaves mostly needlelike.

'Pfitzeriana Nana'   Growth dwarf, rounded; leaves green.

'Pfitzeriana Variegata'   Similar to 'Pfitzeriana'; foliage patched
   yellow.

'Plumosa'   Growth dwarf, often one-sided; sprays crowded; some leaves
   needlelike.

'Plumosa Albovariegata'   Growth dwarf; leaves white at the tips
   of the twigs.

'Plumosa Aurea'   Growth dwarf; branches more on one side; foliage
   with some yellow.

'Plumosa Aureovariegata' Growth dwarf; leaves with yellow patches, some leaves needlelike.

'Procumbens'

'Pyramidalis' Growth columnar, dense; leaves mostly needlelike.

'Pyramidalis Nana'

'Reevesiana'

'Richeson' Growth low, compact; some leaves needlelike.

'Robusta Green' Growth slow, upright; leaves green and tufted on branches.

'Rockery Gem' Same as *J. sabina* 'Rockery Gem'.

'San Jose' Growth dwarf, close to the ground; leaves mostly needlelike.

'Sea Green' Leaves bright green.

'Sea Spray' Growth dwarf, very dense, low.

'Sheppardii' Growth conical.

'Shimpaku' Growth low, irregular in shape.

'Shimpaku Nana'

'Shoosmith' Growth dwarf, rounded, compact; leaves needlelike.

'Smithii' Growth conical to columnar; branches few.

'Spartan' Growth rapid, conical, dense.

'Spearmint' Growth upright; foliage bright green.

'Story' Foliage dark green.

'Stricta' Growth conical, dense; leaves needlelike but soft.

'Sulphur Spray' Shrub; some leaves needlelike; foliage yellow.

'Sylvestris' Growth conical; foliage gray-green.

'Sylvestris Plumosa'

'Titlis' Growth dwarf, columnar, compact, irregular.

'Torulosa' Shrub or tree; growth rapid; twigs tufted, cordlike.

'Torulosa Variegata' Growth slow, upright; foliage spotted with yellow or tipped white.

'Tremonia' Similar to 'Blaauw' but with coloring of 'Plumosa Aurea'.

'Variegata' Growth conical; leaves mostly needlelike; white or spotted with white.

'Variegated Kaizuka' Growth conical, narrow; twigs in groups at ends of branches; leaves with yellow patches, often needlelike.

'Veroides'

'Watereri'

'Wilsonii' Branches crowded together.

'Wilson's Weeping' Growth compact, not weeping; some leaves needlelike.

'Wintergreen' Growth conical; foliage dense.

Juniperus chinensis Cultivar Character Groupings

Dwarf

'Dropmore'   'Kohankie's Compact'   'Plumosa
'Echiniformis'   'Mathot'   Aureovariegata'
'Globosa'   'Pfitzeriana Nana'   'Rockery Gem'
'Japonica'   'Plumosa'   'San Jose'
'Japonica Variegata'   'Plumosa Albovariegata'   'Sea Spray'
'Kallays Compact'   'Plumosa Aurea'   'Shoosmith'
                                                   'Titlis'

Columnar

'Columnaris'   'Hetzii Columnaris'   'Olympia'
'Columnaris Glauca'   'Jacobiana'   'Pyramidalis'
'Excelsior'   'Leeana'   'Titlis'
'Fortunei'   'Mission Spire'
'Glauca'   'Obelisk'

Yellow

'Albospica'   'Kuriwao Gold'   'Pfitzeriana Variegata'
'Aurea'   'Mordigan'   'Plumosa Aurea'
'Aurea Globosa'   'Old Gold'   'Plumosa Aureovariegata'
'Gold Coast'   'Pfitzer Blue Gold'   'Sulphur Spray'
'Japonica   'Pfitzeriana Aurea'
   Aureovariegata'   'Pfitzer Mordigan'

*+2.   *J. communis* L.   Common juniper.**

A native of Europe and Asia as well as North America. Usually a shrub but has many cultivated varieties. Leaves needlelike and joining the twig at one place rather than fusing to the twig for some distance along it. Leaves concave, with white band on the upper surface, usually broader than the green edges.   ssp. *hemisphaerica* Nyman. Shrub; growth rounded, dense, low.   var. *depressa* Pursh. Prostrate juniper. Growth low, branches resting on ground; lower surface of leaves white. Three cultivars of this botanical variety: 'Aurea', leaves yellow becoming green; 'Aureospica', leaves at twig tips yellow; 'Effusa', growth spreading.   var. *saxatilis* Pall. Branches resting on ground, forming a mat.

'Arnold'   Growth columnar.
'Ashfordii'   Growth columnar.

'Aurea' Leaves yellow, becoming green.

'Bakony' Growth dwarf, conical, compact.

'Berkshire' Growth dwarf, spreading or mounded; foliage light blue.

'Blue Pacific' Growth spreading, rapid.

'Bruns' Growth erect, loosely branched.

'Candelabrica' Growth conical; branches spreading, turning up at the ends.

'Candelabriformis' Growth compact; branches drooping somewhat.

'Columnaris' Growth columnar.

'Compacta' Growth dense.

'Compressa' Growth dwarf, columnar, compact; leaves short.

'Conspicua' Growth upright; leaves far apart.

'Controversa' Growth columnar, compact.

'Cracovica' Growth columnar, dense.

'Depressed Star' Growth dwarf, vase-shaped; twigs few.

'Derrynane' Similar to 'Repanda'.

'Dumosa' Growth dwarf; twigs quadrangular.

'Echiniformis' Growth dwarf, compact; leaves small, deep green, prickly.

'Edgbaston' Similar to 'Hornibrookii'.

'Effusa' Growth slow.

'Ellis' Growth dwarf; leaves with some yellow coloring.

'Erecta' Growth columnar; similar to 'Hibernica'.

'Gew Graze' Similar to 'Repanda'.

'Gimborn' Growth low, matlike, dense; leaves crowded.

'Gold Beach' Growth dwarf; branches lying on ground; leaves yellow at branch tips.

'Graciosa' Growth wide spreading; branches thin; leaves thin.

'Grayii' Growth columnar, narrow; upper surface of leaves silver and turned outward.

'Hibernica' Growth columnar, dense; twigs vertical.

'Hornibrookii' Growth low, matlike.

'Hornibrook's Gold'

'Inverleith' Similar to 'Hornibrookii'.

'Kiyonoi' Growth columnar; leaves dark green; similar to 'Hibernica'.

'Laxa' Growth columnar.

'Meyer' Leaves silver green.

'Minima' Growth dwarf, matlike; twigs very short.

'Nana' Growth dwarf.

'Nana Aurea' Growth dwarf; branches spreading on ground; leaves yellow.

'Oblonga Pendula'   Growth erect; twigs drooping.

'Pencil Point'   Growth slow, conical.

'Pendula'   Twigs drooping.

'Pendula Aurea'   Growth conical; twigs drooping, with yellow leaves at tips.

'Prostrata'   Growth low, creeping, close to the ground.

'Pyramidalis'   Growth conical.

'Repanda'   Growth dwarf, matlike.

'Sentinel'   Growth columnar, almost pointed.

'Soapstone'   Similar to 'Repanda'.

'Suecica'   Growth columnar; twigs drooping at tips.

'Suecica Aurea'   Foliage partly yellow.

'Suecica Nana'   Growth dwarf, columnar to conical.

'Vase'   Growth dwarf.

'Windsor Gem'   Similar to 'Minima'.

'Zeal'   Growth dwarf, upright.

Juniperus communis Cultivar Character Groupings

Dwarf

'Bakony'	'Dumosa'	'Gold Beach'	'Soapstone'
'Berkshire'	'Echiniformis'	'Minima'	'Suecica Nana'
'Compressa'	'Effusa'	'Nana'	'Vase'
'Depressed Star'	'Ellis'	'Nana Aurea'	'Windsor Gem'
'Derrynana'	'Gew Graze'	'Repanda'	'Zeal'

Low, Spreading over the Ground

'Derrynana'	'Gold Beach'	'Prostrata'	'Windsor Gem'
'Edgbaston'	'Hornibrookii'	'Repanda'	
'Gew Graze'	'Inverleith'	'Soapstone'	
'Gimborn'	'Minima'	'Suecica'	

Columnar

'Arnold'	'Controversa'	'Hibernica'	'Suecica'
'Ashfordii'	'Cracovia'	'Kiyonoi'	'Suecica Nana'
'Columnaris'	'Erecta'	'Laxa'	
'Compressa'	'Grayii'	'Sentinel'	

3. *J. conferta* **Parl.** **Shore juniper.**

Native to Japan. A low, spreading plant often used as ground cover for sand dunes and increasingly popular elsewhere. Leaves needlelike, crowded, grooved, with a narrow, white longitudinal band on upper surface.

'Blue Pacific' Growth low to ground.
'Boulevard' Main branches horizontal.
'Emerald Green' Growth low.
'Emerald Sea' Growth low, matlike; leaves in threes.

* + 4. *J. horizontalis* **Moench.** **Creeping juniper.**

Native of northern North America. Much employed for a ground cover because of its habit of spreading close to the ground. Can be as tall as 1.3 m (4 ft). Leaves mostly scalelike, bluish green, and emitting an agreeable odor when crushed.

'Admirabilis' Growth dwarf; some leaves needlelike.
'Adpressa' Growth dwarf, low, dense, spreading.
'Alpina' Ends of twigs turning upward; most leaves needlelike.
'Andorra' Same as 'Plumosa'.
'Andorra Compact' Growth rapid, low, rounded, spreading, compact.
'Argentea' Growth matlike; branches crowded.
'Aunt Jemima' Growth dense.
'Aurea' Young leaves yellow.
'Bar Harbor' Growth rapid; stems creeping, following the ground; side branches short, erect.
'Blue Acres' Young foliage very blue.
'Blue Chip' Growth pressed to the ground.
'Blue Horizon' Growth low, open; foliage blue.
'Blue Mat' Growth slow.
'Blue Moon' Growth matlike; branches fine; similar to 'Blue Chip'.
'Blue Rug' Same as 'Wiltonii'.
'Blue Wilton' Same as 'Wiltonii'.
'Bonin Island'
'Coast of Maine' Similar to 'Gray Carpet'.

'Depressa Plumosa'   Same as 'Plumosa'.

'Douglasii'   Growth rapid; some or all leaves needlelike, glaucous, blue.

'Dunvegan Blue'   Foliage glaucous.

'Dwarf Blue'

'Emerald Spreader'   Growth weak; branches crowded, feathery.

'Emerson'   Growth dwarf, dense; some leaves needlelike.

'Eximia'   Growth more upright.

'Filicina'   Growth flat; sprays upright, directed forward.

'Filicina Minima'   Growth dwarf.

'Fountain'   Similar to 'Aunt Jemima'.

'Glauca'   Growth slow; some or all leaves needlelike, glaucous, blue.

'Glenmore'   Branches nearly upright, thin; leaves dark green.

'Glomerata'   Growth dwarf, low, spreading; branches shorter, erect; twigs crowded, tufted; some leaves needlelike.

'Gray Carpet'   Growth cushionlike; leaves green.

'Gray Pearl'   Growth compact.

'Green Acres'   Foliage dark green.

'Hermit'   Growth vigorous, dense, spreading.

'Hughes'   Growth close to ground, spreading; foliage glaucous.

'Humilis'   Branches creeping along ground; twigs nearly erect.

'Jade Spreader'   Growth matlike; leaves jade green.

'Livida'   Growth matlike; branches yellow-brown; leaves needlelike.

'Livingston'   Growth dense, much branched.

'Marcellus'   Growth spreading, low; leaves needlelike.

'Morton Arboretum'   Growth more rapid; branches long, flat, curving up at the ends.

'Petraea'   Growth strong, low.

'Planifolia'   Branches long, strong, fast growing.

'Plumosa'   Growth low, rapid; branches spreading horizontally, leaves all needlelike.

'Plumosa Compacta'   Same as 'Andorra Compact'.

'Prince of Wales'   Growth low, forming dense mat.

'Prostrata'   Growth low and spreading.

'Pulchella'   Growth slow, compact.

'Repens'   Branches weak; leaves gray.

'Schmidtii'

'Sea Spray'   Same as *J. chinensis* 'Sea Spray'.

'Sun Spot'   Growth low, spreading over the ground; foliage with yellow patches.

'Turquoise Spreader' Growth spreading, flat, not mounded; branches dense, crowded.
'Variegata' Twigs erect, tips white; leaves needlelike.
'Venusta' Similar to 'Glauca'.
'Viridis' Branches dense; leaves mostly needlelike.
'Wapiti' Growth dense.
'Watnong'
'Webberi' Growth low, spreading, dense.
'Wilms' Growth low, spreading, compact.
'Wiltonii' Growth dwarf, dense; some or all leaves needlelike.
'Winter Blue' Foliage bright blue in winter.
'Youngstown' Growth close to the ground; leaves green; similar to 'Andorra Compact'.
'Yukon Belle' Growth cushionlike.

5. *J. procumbens* (Endl.) Miq. in Sieb. & Zucc. Creeping juniper.

Native to Japan. Low shrub to 70 cm (2 ft) high, with stiff vertical twigs and branches. Leaves in threes, glaucous, with two white spots near the base.

'Bonin Isles' Growth matlike; branches and foliage crowded.
'Golden' Spray tips yellow.
'Nana' Growth densely matted; leaves smaller.
'Nana Glauca' Growth dwarf.
'Santarosa' Growth dwarf.

6. *J. rigida* Sieb & Zucc. Needle juniper.

Spreading shrub or small tree of southeast Asia. Leaves needlelike, joined to the twig at one position, deeply grooved on upper surface so that the narrow white longitudinal band there is completely hidden.

7. *J. sabina* L. Savin.

Native to Europe and western Asia. Spreading shrub to 4 m (12 ft) tall. Leaves mostly scalelike or needlelike, with a disagreeable odor.

'Albovariegata' Foliage with white patches.

'Arcadia' Growth dwarf; leaves mostly scalelike; similar to 'Tamariscifolia'.

'Aureovariegata' Similar to 'Cupressifolia'.

'Blue Danube' Twigs crowded; leaves scalelike and needlelike.

'Blue Forest' Growth dwarf; spreading.

'Broadmoor' Growth mounded; twigs spreading upward; similar to 'Tamariscifolia'.

'Buffalo' Plant very hardy; leaves green; similar to 'Tamariscifolia'.

'Calgary Carpet' Growth low, dense.

'Cupressifolia' Growth low; some leaves needlelike.

'Erecta' Branches thin.

'Fastigiata' Growth columnar.

'Femina' Similar to 'Cupressifolia'.

'Glauca' Foliage glaucous.

'Hicksii' Branches upright, then spreading.

'Holmbury Hill' Similar to 'Cupressifolia'.

'Humilis'

'Jade' Growth slow; leaves thin.

'Knap Hill'

'Mas' Branches upright then spreading; leaves mostly needlelike.

'Musgrave' Similar to 'Cupressifolia'.

'New Blue' Growth spreading, mounded; leaves mostly needlelike.

'Pepin'

'Rockery Gem' Growth dwarf; main branches horizontal, twigs numerous.

'Skandia' Growth low, spreading; foliage feathery.

'Tamariscifolia' Growth low, rapid; twigs short; leaves mostly needlelike, short.

'Thomsen' Growth dwarf, compact; foliage dark green.

'Variegata' Growth dwarf; branches spreading; twig tips turning down, white.

'Von Ehren' Growth vase-shaped; leaves needlelike.

8. *J. scopulorum* Sarg. Rocky Mountain juniper.

Tree to 12 m (40 ft) tall, native to the Rocky Mountains. Popular for landscape plantings because of the conical shape, silver-blue leaves, and hardiness of the cultivars. Leaves scalelike, opposite, pressed tightly to the twig.

'Admiral'   Growth conical.

'Alba'   Growth conical; leaves blue.

'Argentea'   Growth conical.

'Blue Haven'   Same as 'Blue Heaven'.

'Blue Heaven'   Growth slow, narrow, symmetrical, compact.

'Blue Moon'   Growth slow, compact, dense.

'Blue Sierra'   Growth conical, wide.

'Chandleri'   Growth conical, compact.

'Chandler's Blue'

'Chandler's Silver'   Growth conical; leaves needlelike.

'Cologreen'   Growth columnar.

'Columnaris'   Growth columnar.

'Columnar Sneed'   Growth columnar; foliage compact.

'Commando'   Growth conical.

'Cupressifolia'   Growth conical, dense; twigs somewhat drooping.

'Cupressifolia Erecta'   Growth upright, tall.

'Cupressifolia Glauca'   Growth conical; twigs drooping; leaves
   glaucous.

'Dew Drop'   Growth columnar, compact, dense.

'Erecta'   Growth columnar; some leaves needlelike.

'Erecta Glauca'

'Fainii'   Foliage soft and feathery.

'Funalis'   Branches slender.

'Gareei'   Growth dwarf, rounded, compact.

'Glauca'   Growth compact; leaves glaucous.

'Globe'   Growth rounded; foliage feathery.

'Gracilis'   Growth compact; leaves more green.

'Gray Gleam'   Growth conical, dense; branches crowded; leaves gray.

'Greenspire'   Growth columnar.

'Hall's Sport'   Growth compact, upright.

'Hillborn's Silver Globe'   Growth dwarf, rounded.

'Hill's Silver'   Growth columnar.

'Holmes Silver'   Leaves silver.

'Kansas Silver'   Growth conical; ends of twigs erect.

'Kenyonii'   Growth slower, compact.

'Lakewood Globe'   Growth rounded.

'Medora'   Growth columnar; foliage blue.

'Moffet Blue'   Growth columnar, symmetrical, upright; foliage intense
   blue.

'Moffetii'   Growth conical; foliage dense.

'Montana'   Growth slow; leaves dark green.

'Moonglow'   Growth upright, compact.

'Moonlight'   Leaves silver blue.

'Mounteneer'   Growth conical.

'North Star'   Growth compact.

'O'Conner'   Branches crowded, many.

'Palmeri'   Growth compact, low to the ground; leaves needlelike.

'Pathfinder'   Growth conical.

'Pendula'   Branches drooping.

'Platinum'   Growth conical, dense.

'Prostrata'   Growth low, spreading.

'Repens'   Growth dwarf; leaves needlelike.

'Salome's Blue'

'Silver Beauty'   Growth conical.

'Silver Cord'   Growth thin, weak.

'Silver Glow'   Growth conical; foliage glaucous.

'Silver King'   Growth dwarf; twigs often cordlike.

'Silver Queen'   Leaves needlelike and scalelike; very glaucous.

'Silver Star'   Growth wide, low.

'Skyrocket'   Growth extremely narrow but branches spreading with age.

'Springbank'   Growth slow, conical.

'Steel Blue'   Growth weak.

'Sutherland'   Growth dense; branches erect.

'Tabletop'   Growth dwarf; twigs thin; leaves small.

'Tabletop Blue'   Growth dense, spreading.

'Tolleson's Weeping'   Growth conical; branches crowded; twigs drooping.

'Viridifolia'   Growth narrow; leaves bright green.

'Welchii'   Growth columnar.

'White's Silver King'

'Wichita Blue'   Growth conical; foliage blue.

9.   *J. squamata* Lamb.   Himalayan juniper.

Native to Asia. Shrub with thick twigs and dense foliage. Leaves needlelike and in threes.

'Blue Carpet'   Growth low, creeping; foliage similar to 'Meyeri'.

'Blue Spider'

'Blue Star'   Growth dwarf, dense, irregular, mounded; foliage steel blue.

'Chinese Silver'   Growth wide, spreading; branches drooping.

'Forrestii'   Similar to 'Wilsonii'.

'Glassel'   Growth slow; branches at steep angles; twigs curved.

'Golden Flame'   Similar to 'Meyeri' but leaves spotted yellow.

'Holger'   Growth spreading; leaves on new shoots yellow.

'Loderi'   Growth conical with multiple leading shoots; branches and twigs very crowded.

'Meyeri'   Branches erect; twigs crowded, short, straight; leaves crowded.

'Parsonii'   Growth low, wider than high, vaselike.

'Prostrata'   Growth dwarf; stems and branches long and creeping.

'Pygmaea'   Similar to 'Prostrata'.

'Wilsonii'   Branches crowded, recurved; leaves shorter.

*+10.   *J. virginiana* L.   Eastern red-cedar.

Tree to 33 m (100 ft). Native to eastern North America. Cultivated in many forms. Leaves scalelike or needlelike, acute, opposite on mature branches, and lacking strong odor.

'Albospica'   Some leaves white at twig tips.

'Aurea'   Foliage yellow.

'Blue Cloud'   Same as *J. chinensis* 'Blue Cloud'.

'Boskoop Purple'   Growth rapid, columnar, dense.

'Burkii'   Growth conical, leaves glaucous, purplish in winter.

'Burkii Compacta'   Growth conical, dense, compact; foliage steel blue with purple tinge in winter.

'Canaertii'   Growth columnar to conical; twigs cordlike, clustered.

'Chamberlaynii'   Shrub; branches spreading, reflexed; leaves mostly needlelike.

'Cinerascens'   Growth conical; young leaves gray.

'Columnaris'   Growth columnar.

'Cupressifolia'   Growth conical; leaves yellow-green.

'DeForest Green'   Growth conical.

'Elegantissima'   Growth conical; twig tips yellow.

'Emerald Sentinel'   Growth rapid, columnar.

'Fastigiata'   Growth columnar.

'Filifera'   Growth conical.

'Glauca'   Growth columnar; foliage very glaucous.

'Glauca Compacta'   Growth columnar; foliage glaucous.

'Glauca Hetzii'   Growth dense, spreading; foliage feathery, glaucous.

'Glenn Dale'

'Globosa'   Growth compact, rounded; branches numerous, crowded; twigs short, thin.

'Gray Owl'   Growth spreading, rapid; twigs cordlike.

'Green Spreader'   Growth spreading, low; foliage threadlike.

'Helle'   Growth compact, upright.

'Henryi'   Growth conical.

'Hillii'   Growth columnar, dense; foliage turns reddish in winter.

'Hillspire'   Growth conical; leaves bright green.

'Humilis'   Growth dwarf, spreading; leaves in tufts.

'Keteleeri'   Growth conical, rapid.

'Kobenzeii'   Growth slender, branches many; some leaves needlelike.

'Kobold'   Growth dwarf, conical; branches thin, crowded, many.

'Kosteri'   Growth rounded, dense; branch tips feathery; some leaves needlelike.

'Lebretonii'

'Manhattan Blue'   Growth conical, compact.

'McCabei'

'Mission Spire'   Growth upright.

'Nana'   Growth dwarf.

'Nana Compacta'   Growth dwarf; foliage tinged purple in winter.

'Nevin's Blue'

'Nova'   Growth columnar, upright.

'North Star'   Same as *J. scopulorum* 'North Star'.

'Pendula'   Growth low; open; twigs drooping.

'Pendula Nana'   Growth dwarf; branches spreading.

'Pendula Viridis'   Twigs drooping; foliage bright green.

'Plumosa'   Growth conical.

'Plumosa Argentea'   Branches few, twigs short.

'Prostrata'   Growth low, spreading.

'Pseudocupressus'   Growth columnar, branches erect.

'Pumila'   Growth dwarf, rounded, compact; twigs crowded; some leaves needlelike.

'Pyramidalis'   Growth columnar.

'Pyramidiformis'   Growth columnar; leaves dark green.

'Ramlosa'   Growth spreading, vase-shaped; foliage feathery.

'Reptans'   Growth dwarf, low, flat; twig tips drooping; some leaves needlelike.

'Robusta Green'   Growth columnar; branches crowded; most leaves needlelike.

'Schottii'   Growth conical, dense.

'Sherwoodii'   Growth conical; shoot tips dark red in winter.

'Silver Spreader'   Growth spreading, low; leaves gray.

'Skyrocket'   Same as *J. scopulorum* 'Skyrocket'.

'Smithii'

'Sparkling Skyrocket'   Growth extremely narrow; foliage patched with amber.

'Staver Blue'   Growth conical; foliage silver-blue.

'Tripartita'   Shrub; leaves mostly needlelike.

'Venusta'   Growth conical; leaves gray.

Larix kaempferi

Larix laricina

Larix Mill.   Larch.

Larix, a genus of 15 species, occurring in the northern regions and high altitudes of the Northern Hemisphere, is easy to recognize. The leaves are clustered in groups of 12–50. They become golden and drop in autumn. The twigs are irregularly spaced on the trunk, and branches bear short knobs or spurs that are actually extremely slow growing, condensed branches. The cones are erect and short stalked, 1.3–4.0 cm (½–1¾ in) long, with bracts that protrude at first but are often hidden when mature. Larches thrive with a short growing season and maximum light. Specimen trees need much space, as they produce more horizontal growth than most conifers. The species of larch are difficult to distinguish without the aid of cone characters.

83. *L. kaempferi*

1. Leaves with 2 noticeable white longitudinal bands on the lower
     surface (Fig. 83).
    2. Young shoots brown, glaucous; leaves deeply keeled, the
       bands wider on the lower surface .... *L. kaempferi*
                                                         Japanese larch

2. Young shoots red-brown, somewhat glaucous; leaves not as deeply keeled, the bands narrower on the lower surface .... *L.* ×*eurolepis*
Dunkeld larch

1. Leaves with 2 inconspicuous pale longitudinal bands on the lower surface (Fig. 84).

84. *L. decidua*

3. Leaves mostly less than 3.3 cm (1¼ in) long, angled (Fig. 85) .... *L. laricina*⁺
American larch

85. *L. laricina*

3. Leaves mostly greater than 3.3 cm (1¼ in) long, upper surface flat (Fig. 86).

86. *L.* ×*eurolepis*

4. Leaves mostly less than 3.7 cm (1½ in) long.

5. Twigs glaucous .... *L.* ×*eurolepis*
Dunkeld larch

5. Twigs not glaucous .... *L. decidua*
European larch

4. Leaves mostly greater than 3.7 cm (1½ in) long ..... *L. sibirica*
Siberian larch

1. *L. decidua* Mill.   European larch.

Large tree from the mountains of Europe and Siberia. Leaves 1–3 cm (⅜–1⅛ in) long. Cones 2–4 cm (¾–1½ in) long.

'Compacta'   Growth conical, dense; branches brittle.
'Conica'   Growth conical; branches horizontal, then turning downward.
'Corley'   Growth conical, not so rapid.
'Fastigiata'   Growth columnar.
'Pendula'   Branches deeply arching.
'Pyramidalis'   Growth columnar; similar to 'Fastigiata'.
'Repens'   Growth dwarf; lower branches lying on ground.
'Tortuosa'   Branches twisted.
'Viminalis'   Branches drooping.
'Virgata'   Branches few, long, snakelike.

2. *L.* ×*eurolepis* A. Henry (*L. kaempferi* Carr. × *L. decidua* Mill.)   Dunkeld larch.

A chance hybrid of vigorous growth. Difficult to distinguish from the parent species, *L. decidua* and *L. kaempferi*. Distinguished from the former by its wider, glaucous leaves and slightly turned-back cone scales and bracts and from the latter by less glaucous, yellow twigs and shorter leaves with less prominent white bands. Leaves 2–4 cm (¾–1½ in) long. Cones 2–4 cm (¾–1½ in) long.

3. *L. kaempferi* (Lamb.) Carr.   Japanese larch.

Native to Japan. The most popular ornamental larch. Rapid growth and tolerance to exposure. Leaves 1.3–4.5 cm (½–1¾ in) long, light blue to blue-gray. Cones 1.3–4.0 cm (½–1½ in) long.

'Blue Rabbit'   Growth conical, slender; leaves glaucous.
'Dervaes'   Branches horizontal; twigs drooping.
'Ganghoferi'   Growth conical, compact.
'Inversa'   Branches drooping.
'Lombarts'   Twigs drooping.
'Minor'   Growth dwarf.
'Nana'   Growth dwarf, congested, compact.
'Pendula'   Growth slow; branches drooping.
'Pyramidalis Argentea'   Branches and twigs bending backward.
'Varley'   Growth dwarf, dense.
'Wehlen'   Growth dwarf; irregularly wide spreading, compact.
'Wolterdingen'   Growth dwarf, irregular, wide.

+4. *L. laricina* K. Koch.   American larch.   Tamarack.

Native to northern North America. Often found in swamps or wet areas. Known also in cultivation. Leaves 2.5–3.3 cm (1–1¼ in) long. Cones 1.3–2.0 cm (½–¾ in) long.

'Alaskensis'
'Depressa'
'Glauca'   Leaves glaucous.

5. *L. sibirica* Ledeb.   Siberian larch.

Introduced from Soviet Union and rarely planted. Leaves 2–4 cm (1–2 in) long. Cones 3–4 cm (1¼–1½ in) long.

Metasequoia glyptostroboides

Metasequoia Miki. Dawn-redwood.

This genus includes only one species and was known only through the fossil record until living plants were found in China in the 1940s.

1. *M. glyptostroboides* Hu & Cheng. Dawn-redwood. Dawn-cypress.

A species only recently discovered growing in a small area in the provinces of Hupeh and Szechuan, China. Cultivated mostly only in arboreta. Makes a pleasing and interesting specimen because of its shredding red bark, feathery, loosely spaced leaves and branches, which drop in the fall. Leaves needlelike, light green, opposite, 1.3–3.8 cm (½–1½ in) long. Cones 2–3 cm (¾–1¼ in) long.

'National' Growth narrowly conical.

Microbiota decussata

Microbiota Kom.　Microbiota.

This genus is relatively new to horticulture. Its only species was first discovered about 60 years ago in southeastern Soviet Union.

1.　*M. decussata* Kom.　Microbiota.

Native to a small area in southeastern Siberia. A dwarf shrub less than 1 m (3 ft) tall, with spreading branches that stay close to the ground. Leaves on inner branches toward the trunk, needlelike, and without glands. Other leaves scalelike, equal in size, 2 mm (¹⁄₁₆ in) long, 1 mm wide, with glands. Female reproductive structure roughly resembling that of the junipers, although 2 or 4 cones scales are present. An excellent choice for ground cover in a rock garden or other area where a low, trailing plant is desired. Not common in cultivation despite its hardiness and the ease with which it may be propagated. May become more popular as these plants become more available.

Picea abies

Picea omorika

Picea A. Dietr. Spruce.

This genus consists of about 35 species of trees in the Northern Hemisphere that are often uniformly conical and grow to great heights. The leaves are angled in cross section, with white longitudinal bands on the upper leaf surface and usually on the lower leaf surface. The leaves are attached to the twig on woody projections from the twig. The cones hang down from the ends of the branches. Most species are easily cultivated in the Northeast. Many cultivars have been selected by rooting and grafting. The genus is a favorite of dwarf conifer growers. Scions from all species are almost always very successfully grafted on *Picea abies* understock.

87. *P. jezoensis*

88. *P. jezoensis*

89. *P. sitchensis*

1. Leaves much wider than high (Fig. 87), with longitudinal white bands on upper surface only.
  2. Twigs (except young shoots) lacking hairs; leaves flat.
    3. Leaf tips abruptly pointed (Fig. 88) . . . . *P. jezoensis*
      Yezo spruce
    3. Leaf tips tapering, sharp (Fig. 89) . . . . *P. sitchensis*
      Sitka spruce
  2. Twigs hairy; leaves slightly angled . . . . *P. omorika*
    Serbian spruce

101

91. *P. mariana*

93. *P. orientalis*

95. *P. orientalis*

97. *P. torano*

99. *P. engelmannii*

101. *P. engelmannii*

90. *P. glauca**

92. *P. mariana*

94. *P. orientalis*

96. *P. asperata*

98. *P. alcoquiana*

100. *P. pungens*

1. Leaves nearly as high as wide (Fig. 90); longitudinal white
   bands on all 4 surfaces.
   4. Twigs (except young shoots) obviously hairy.
      5. Leaves blue, usually greater than 16 mm (⅝ in)
         long . . . . *P. engelmannii*
                                                                 Engelmann spruce
      5. Leaves green, less than 16 mm (⅝ in) long.
         6. Scales at the base of the bud awl-shaped, with long
            tips (Fig. 91); twig hairs usually glandular (Fig. 92) (best
            seen on new growth).
            7. Leaves 6–13 mm (¼–½ in) long, straight,
               glaucous . . . . *P. mariana*⁺
                                                                 Black spruce
            7. Leaves 13–16 mm (½–⅝ in) long, incurved,
               sometimes twisted . . . . *P. rubens*⁺
                                                                 Red spruce
         6. Scales at the base of the bud not usually awl-shaped,
            with rounded or blunt tips (Fig. 93); twig hairs
            lacking glands (Fig. 94).
            8. Leaves blunt or notched (Fig. 95) . . . .*P. orientalis*
                                                                 Oriental spruce
            8. Leaves sharp-pointed or at least with beveled tips
               (Fig. 96).
               9. Leaves greater than 13 mm (½ in) long, tips of lower
                  bud scales rounded, not awl-shaped . . . . *P. asperata*
                                                                 Dragon spruce
               9. Leaves often less than 12 mm (½ in) long; tips of
                  some lower bud scales awl-shaped . . . . *P. glehnii*
                                                                 Sakhalin spruce
   4. Twigs lacking hairs.
      10. Leaves with very sharp, spiny tips (Fig. 97), higher than
          wide in cross section; rare . . . . *P. torano*
                                                                 Tigertail spruce
      10. Leaves only sharp-pointed (Fig. 98), as wide as or wider
          than high in cross section.
          11. Buds large, to 10 mm (⅜ in) long, with scales
              curving away from bud (Fig. 99).
              12. Buds gummy or sticky . . . . *P. asperata*
                                                                 Dragon spruce
          12. Buds dry, not sticky.
              13. Scales at the base of the bud often with long
                  points (Fig. 100); leaves stiff, blue, 6–16 mm
                  (¼–⅝ in) long . . . . *P. pungens**
                                                                 Blue spruce
              13. Scales at the base of the bud rounded, not long-
                  pointed (Fig. 101); leaves more flexible, with strong

102. *P. abies*

odor when crushed, 1.3–2.5 cm
(½–1 in) long . . . . *P. engelmannii*
Engelmann spruce
11. Buds small, with scales pressed tightly together (Fig.
102).
14. Leaves glaucous, with strong odor when crushed, stiff; but
scales often with divided tips . . . . *P. glauca* +
White spruce
14. Leaves not glaucous, with odor not so strong; bud scales
with rounded tips.
15. Leaves spreading at right angles to the twig;
shrubs . . . . *P. maximowiczii*
Japanese bush spruce
15. Leaves pointed forward along twig; trees.
16. Leaves gray on upper surfaces and dark green on
lower surfaces, often curved and stiff . . . . *P. alcoquiana*
Alcock spruce
16. Leaves dark green on all surfaces, straight . . . . *P. abies**
Norway spruce

*1. *P. abies* (L.) Karst.　Norway spruce.

A native of Europe that has been the most commonly planted ornamental conifer in the United States, its graceful, sweeping boughs
on mature trees and rapid growth making it popular in cultivation.
Many cultivars selected for all types of landscape planting, from
dwarf conifers and hedges to specimen trees. Leaves 1.3–2.0 cm
(½–¾ in) long. Cones 10–18 cm (4–7 in) long.

'Aarburg'　Growth irregular; branches many; leaves short.
'Acrocona'　Growth irregular; cones producing early.
'Acutissima'　Growth slow; leaves long, thin, sharp-pointed.
'Araucarioides'　Whorls of branches spaced far apart near top of
tree.
'Argentea'　Leaves with patches of white.
'Argenteospicata'　Leaves on young shoots yellow, then dark green.
'Arnold Dwarf'
'Asselyn'
'Aurea'　Leaves with yellow cast toward tips on upper surfaces.
'Aurea Magnifica'　Growth lower, bushy; leaves yellow.
'Aurescens'　Leaves yellow when young, becoming yellowish green.
'Barnes'　Growth dwarf, rounded.
'Barryi'　Growth conical; branches thick, erect.
'Beissneri'　Growth dwarf; branches thick; twigs often short.

'Bennett's Miniature'   Growth dwarf.

'Bergman's Flat Top'

'Bergman's Striped Leaf'

'Brevifolia'   Leaves shorter.

'Capitata'   Growth dwarf, rounded; young shoots close together, forming heads at least every other year.

'Cellensis'   Growth dwarf, conical; branches fine; leaves short.

'Cincinnata'   Branches drooping, widely spaced at the top.

'Chlorocarpa'   Cones green when young.

'Clanbrassiliana'   Growth slow or even dwarf, rounded, compact.

'Clanbrassiliana Elegans'   Growth slow, conical.

'Clanbrassiliana Plumosa'   Growth slow; leaves twisted near tips.

'Clanbrassiliana Stricta'   Growth slow, conical, very compact, very similar to 'Clanbrassiliana Elegans'.

'Coerulea'   Leaves blue to blue-gray.

'Columnaris'   Growth columnar; branches short.

'Compact Asselyn'   Growth dwarf, conical, very compact.

'Compacta'   Growth dense, rounded.

'Conica'   Growth dwarf, conical.

'Corbit'

'Corticata'   Bark very thick.

'Costickii'   Growth dwarf, upright.

'Cranstonii'   Branches long, few.

'Crippsii'   Growth dwarf, conical, spreading; branches erect.

'Crusita'   Young leaves at twig tips red.

'Cupressina'   Growth columnar; branches ascending; new growth red.

'Decumbens'   Growth dwarf.

'Denudata'   Growth conical, open.

'Dicksonii'   Branches few; similar to 'Cranstonii'.

'Diffusa'   Growth dwarf, dense, spreading; leaves yellowish.

'Doversii Pendula'   Growth dwarf.

'Dumosa'   Growth dwarf, close to the ground.

'Echiniformis'   Growth dwarf, low, cushion-shaped; leaves small, thin, prickly.

'Elegans'   Growth dwarf, conical; branches crowded.

'Elegantissima'   First-year leaves yellow, turning white.

'Ellwangeriana'   Growth irregular, open; twigs crowded.

'Erimita'   Growth conical; branches erect.

'Finedonensis'   Leaves yellow, becoming bronze, then green.

'Formanek'   Growth dwarf; branches drooping.

'Frohburg'   Main branches drooping; side branches weakly drooping.

'Globosa Nana'   Similar to 'Nana'.

'Gracilis'   Growth dwarf.

'Gregoryana'   Growth dwarf, compact but open, cushion-shaped; twigs crowded, short; leaves prickly, otherwise similar to 'Echiniformis'.

'Gregoryana Parsonsii'   Growth dwarf.

'Gregoryana Veitchii'   Growth more rapid than 'Gregoryana'.

'Gymnoclada'   Leaves remain on twigs for one year only.

'Highlandia'   Growth rounded, low; leaves glossy dark green.

'Hillside Dwarf'   Growth dwarf.

'Hillside Upright'   Growth slow, conical, dense; foliage dark green.

'Holmstrup'   Growth dwarf, conical; branches fine.

'Hornibrookii'   Growth dwarf; branches horizontal; twigs flattened.

'Humilis'   Growth dwarf, conical, compact; twigs crowded.

'Humphrey's Gem'

'Hystrix'   Growth dwarf, open; twigs short, stiff; leaves crowded.

'Intermedia'   Branches dissimilar, thin, long, few; leaves dissimilar.

'Inversa'   Growth low, spreading over the ground; branches crowded, drooping.

'Iola'   Branches twisted.

'Kalmthout'   Similar to 'Nidiformis'.

'Kamon'   Growth dwarf, dense; foliage lustrous.

'Kingsville Fluke'   Growth slow, irregular.

'Kluis'   Growth dwarf, irregular; leaves small.

'Knaptonensis'   Growth dwarf, cushion-shaped, wider than high.

'Lincoln'

'Little Gem'   Growth dwarf; twigs turned up.

'Little Joe'   Growth dwarf, conical.

'Loreley'   Growth slow, drooping; stem curved; lower branches creeping along the ground.

'Major'   Growth drooping, robust.

'Mariae Orffiae'   Growth dwarf, rounded; twigs short.

'Maxwellii'   Growth dwarf, rounded, low; branches very short and thick.

'Merkii'   Growth low, strong; twigs short, thick, and of irregular length.

'Microphylla'   Growth dwarf; leaves small.

'Microsperma'   Growth dwarf, conical, dense, and bushy; cones small.

'Minima'

'Minutifolia'   Growth dwarf; leaves small.

'Monstrosa'   Growth unbranched, one main stem; leaves stiff.

'Montnomah'   Growth tending to be dwarf, irregular.

'Moscowensis'

'Mucronata'   Growth dwarf when young, becoming tall, conical; twigs crowded.

'Multonomah'   Same as 'Montnomah'.

'Mutabilis'   Growth slow, cushionlike.

'Nana'   Growth dwarf, variable.

'Nana Compacta'   Growth dwarf, compact.

'Nidiformis'   Growth dwarf, rounded, dense, concave.

'Ohlendorfii'   Growth dwarf, rounded, becoming conical, compact; leaves yellowish.

'Oldhamiana'

'Pachyphylla'   Growth dwarf, squat, later upright; leaves thick, fleshy.

'Parsonsii'   Growth dwarf; twigs drooping.

'Parviformis'   Growth slow, conical.

'Pendula'   Branches drooping; leader upright.

'Pendula Bohemica'   Growth drooping, two-stemmed.

'Pendula Monstrosa'   Similar to 'Major' but main branches more strongly bent downward.

'Phylicoides'   Branches few; leaves few, thickened.

'Procumbens'   Growth dwarf, spreading over the ground; similar to 'Pseudoprostrata'.

'Prostrata'   Growth slow, low, spreading over the ground.

'Pruhoniceana'   Branches thin, long, few; leaves short.

'Pseudomaxwellii'   Growth rounded.

'Pseudoprostrata'   Growth dwarf, spreading over the ground; twigs crowded, irregularly spaced.

'Pumila'   Growth dwarf, dense; leaves blue-green.

'Pumila Glauca'   Growth dwarf, rounded if rooted, conical if grafted, compact; leaves glaucous.

'Pumila Nigra'   Growth dwarf, rounded if rooted, conical if grafted; branches spreading; leaves crowded on the upper side of twigs.

'Pumilio'   Growth dwarf; branches thick; twigs short, many.

'Pygmaea'   Growth dwarf, conical, dense; winter buds dark brown; similar to 'Humilis'.

'Pyramidalis'   Growth slow, conical.

'Pyramidalis Gracilis'   Growth dwarf, rounded.

'Pyramidata'   Growth conical.

'Reflexa'   Growth spreading over the ground, matlike; branches
   drooping.

'Remontii'   Growth slow, conical, dense.

'Repens'   Growth dwarf, spreading over the ground; leaves crowded.

'Rothenhaus'   Leaves short.

'St. James'   Growth dwarf, rounded.

'Sargentii'   Growth dwarf; branches dense, spreading over the ground.

'Sherwood'   Same as 'Sherwood Gem'.

'Sherwood Gem'   Growth rounded, compact.

'Sherwood Multnomah'   Growth dwarf, then rapid; some branches shed
   annually.

'Tabuliformis'   Growth dwarf.

'Tenuifolia'   Twigs few; leaves slender.

'Tuberculata'   Branches swollen at the base.

'Turfosa'   Growth bushy; branches thick, long.

'Variegata'   Foliage with patches of yellow.

'Veitchii'   Growth dwarf, conical.

'Veitchii Nana'

'Verkades Dwarf'

'Viminalis'   Branches horizontal, long, spreading, curved down; twigs
   slender; leaves shorter toward twig tips.

'Virgata'   Branches few, long and snakelike, often without twigs.

'Wagneri'   Growth dwarf, rounded, compact.

'Wansdyke Miniature'   Growth dwarf; stem short, thick.

'Wartburg'   Branches horizontal or bending close to ground.

'Waugh'   Growth dwarf, open; twigs thick; leaves widely spaced.

'Wells Green Globe'   Growth slow, rounded.

'Wills Zwerg'   Growth dwarf, conical.

'Wilson'   Growth dwarf.

Picea abies Cultivar Character Groupings

Dwarf

'Barnes'	'Crippsii'	'Echiniformis'
'Capitata'	'Diffusa'	'Elegans'
'Compact Asselyn'	'Doversii Pendula'	'Formanek'
'Costickii'	'Dumosa'	'Gracilis'
'Gregoryana'	'Maxwellii'	'Pumila Glauca'

'Gregoryana Parsonsii' 'Microphylla' 'Pumila Nigra'
'Gregoryana Veitchii' 'Microsperma' 'Pumilio'
'Hillside Dwarf' 'Minutifolia' 'Pygmaea'
'Hornibrookii' 'Nana' 'Pyramidalis Gracilis'
'Humilis' 'Nana Compacta' 'Repens'
'Hystrix' 'Nidiformis' 'St. James'
'Kalmthout' 'Ohlendorfii' 'Sargentii'
'Kamon' 'Pachyphylla' 'Tabuliformis'
'Kluis' 'Parsonsii' 'Veitchii'
'Little Gem' 'Procumbens' 'Wagneri'
'Little Joe' 'Pseudoprostrata' 'Waugh'
'Mariae Orffiae' 'Pumila' 'Wilson'

Rounded

'Barnes' 'Kalmthout' 'Pseudomaxwellii'
'Capitata' 'Little Gem' 'Pumila Glauca'
'Clanbrassiliana' 'Mariae Orffiae' 'Pumila Nigra'
'Compacta' 'Maxwellii' 'Pyramidalis Gracilis'
'Echiniformis' 'Mutabilis' 'St. James'
'Gregoryana' 'Nidiformis' 'Wagneri'
'Highlandia' 'Ohlendorfii' 'Wells Green Globe'

2. *P. alcoquiana* (J. G. Veitch ex Lindl.) Carr.   Alcock Spruce.

Formerly known as *Picea bicolor*. A hardy native of Japan with a pleasing, symmetrical growth habit. Foliage often silvery in appearance. Leaves 1.3–2.0 cm (½–¾ in) long. Cones 8–10 cm (3–4 in) long.   var. *acicularis* Shiras & Koyama. Twigs fine, hairy; leaves crowded, narrow, curved.   var. *reflexa* Shiras & Koyama. Twigs hairy; buds lacking resin; leaves shorter.

'Howell's Dwarf'   Growth dwarf, wider than high when young, becoming higher than wide; branches arching; leaves with light blue lower surfaces.
'Prostrata'   Growth low, spreading.

3. *P. asperata* Mast.   Dragon spruce.   Chinese spruce.

Native to western China. Bushy and slow growing at first, rapid growing later. Bark flaky. Leaves swollen, 1.3–2.0 cm (½–¾ in) long. Cones 8–10 cm (3–4 in) long.   var. *notabilis* Rehd. & Wils. Cones larger, 9–14 cm (3½–5½ in) long.

'Hunnewelliana'   Growth dwarf, cushion-shaped, dense; branches short, erect.

4.   *P. engelmannii* Parry ex Engelm.   Engelmann spruce.

Native to the mountains of western North America. An occasionally planted, slow-growing ornamental. Buds resinous. Leaves stiff, blue-gray, 1.3–2.5 cm (½–1 in) long. Cones 4–8 cm (1½–3 in) long.

'Argentea'   Leaves silver.
'Compact'   Growth dwarf.
'Fendleri'   Twigs drooping; leaves long, slender.
'Glauca'   Leaves glaucous.
'Microphylla'   Growth dwarf, compact; leaves smaller.
'Pendula'   Twigs drooping.
'Snake'   Growth open; branches long and few.
'Vanderwolf's Blue Pyramid'   Growth dense; branches ascending.

*+5.   *P. glauca* (Moench.) Voss.   White spruce.

A native northeastern North American species that grows rapidly. Often used as a Christmas tree, although its leaves dry and drop quickly like those of most spruces. Used to a lesser extent as an ornamental or hedge. Leaves blue-green, sometimes quite stiff, when bruised emit a strong odor that can be considered disagreeable, 1–2 cm (⅓–¾ in) long. Cones 4–5 cm (1½–2 in) long.
var. *albertiana* (S. Brown) Sarg. Growth conical, very compact; leaves stiff, 1.2–2.5 cm (½–1 in) long.

'Alberta Globe'   Similar to 'Conica'.
'Aurea'   New growth yellow.
'Brevifolia'   Growth slow; leaves short.
'Cecilia'   Growth dwarf; foliage dense.
'Coerulea'   Leaves crowded, short, silver.
'Compacta'   Growth dwarf.

'Conica'   Growth dwarf, conical, very dense; twigs and leaves spaced close together; leaves thin, light glaucous green.

'Cy's Wonder'   Growth dwarf, open.

'Densata'   Growth slow, symmetrical.

'Dent'   Some leaves yellow.

'Echiniformis'   Growth dwarf, cushion-shaped, dense; twigs often five together.

'Elegans Compacta'   Similar to 'Conica'.

'Ericoides'   Similar to 'Echiniformis' but growth more rapid.

'Fort Ann'   Growth rapid; branches few, twisted.

'Gnom'   Similar to 'Conica'.

'Gnome'   Growth dwarf.

'Gracilis Compacta'   Similar to 'Conica'.

'Hendersonii'   Growth dense but drooping; similar to 'Coerulea'.

'Hillside'   Growth dwarf, rounded, higher than wide.

'Hudsonii'   Leaves blue.

'Laurin'   Similar to 'Conica'.

'Lilliput'   Similar to 'Conica'.

'Little Globe'   Growth slow, rounded.

'Millstream Broom'   Growth dwarf.

'Monstrosa Nana'   Growth dwarf, conical; leaves short.

'Nana'   Growth dwarf, rounded, dense; twigs crowded.

'Pendula'   Branches drooping, many; bark thick.

'Pixie'   Growth dwarf.

'Sander's Blue'   Growth dwarf; leaves glaucous.

'Tabuliformis'   Growth dwarf; branches horizontal; leaves crowded.

'Tiny'   Growth dwarf, tight.

'Wild Acres'   Growth slow.

Picea glauca Cultivar Character Grouping

Dwarf

'Alberta Globe'	'Gnome'	'Nana'
'Cecilia'	'Hillside'	'Pixie'
'Compacta'	'Laurin'	'Sander's Blue'
'Conica'	'Lilliput'	'Tabuliformis'
'Cy's Wonder'	'Millstream Broom'	'Tiny'
'Echiniformis'	'Monstrosa Nana'	

6. *P. glehnii* Mast.   Sakhalin spruce.

Native to Japan and Soviet Union (Sakhalin Island). A short-leaved, hardy spruce with excellent ornamental quality. Leaves 8–15 mm (¼–½ in) long. Cones 5–8 cm (2–3 in) long.

7. *P. jezoensis* (Sieb. & Zucc.) Carr.   Yezo spruce.

Native to northeast Asia and Japan. Rare in cultivation. Flat leaves with white bands only on upper surface, unlike most spruces. Leaves 1.3–2.0 cm (½–¾ in) long. Cones 4–9 cm (1½–3½ in) long.   var. *hondoensis* (Mary) Rehd. Leaves shorter, blunt, silver-blue on lower surface.

'Aurea'   Leaves on upper side of twigs tending to be yellow.
'Nana'   Growth dwarf.

+8. *P. mariana* (Mill.) BSP.   Black spruce.

A native of northern North America that is not cultivated because of its sparse foliage and preference for swamps or wet areas. Leaves 0.7–2.0 cm (¼–¾ in) long. Cones 1.3–4.0 cm (½–1½ in) long.

'Aurea'   Leaves yellow.
'Beissneri'   Growth conical, dense; twigs thin; leaves coarse, steel blue.
'Beissneri Compacta'   Growth conical, compact; leaves coarse.
'Corbit'
'Doumettii'   Growth dwarf, conical, dense; leaves crowded.
'Empetroides'   Growth dwarf.
'Ericoides'   Growth slow, conical; leaves thin.
'Fastigiata'   Growth columnar.
'Globosa'   Growth rounded, dense, dwarf.
'Golden'   Growth slow; new foliage yellow.
'Nana'   Growth dwarf, rounded; leaves blue-green.
'Pendula'   Branches drooping.
'Procumbens'   Growth dwarf, spreading; branches close to the ground.
'Pygmaea'   Similar to 'Nana' but smaller.
'Semiprostrata'   Growth dwarf.

9. *P. maximowiczii* Reg.   Japanese bush spruce.

Native to Japan, where it can be a large tree. In cultivation a small, compact tree or shrub. Leaves stiff, 9–15 mm (⅓–⅗ in) long. Cones 4.0–6.5 cm (1½–2½ in) long.

10. *P. omorika* Purkyne.   Serbian spruce.   Omorika spruce.

Native to Asia. Another of the spruces with white longitudinal bands only on the upper leaf surface, but its leaves are slightly angled. Grows very well as an ornamental. Large planting area unnecessary, as this species grows straight, tall, and narrow. Branches drooping and becoming nearly parallel to the trunk as the tree ages. Leaves a soft, gray-green, 9–13 mm (⅓–½ in) long. Cones 4–6 cm (1½–2½ in) long.

'Alpestris' Twigs densely hairy.

'Aurea' Leaves yellow.

'Berliners Weeping' Probably the same as 'Expansa'.

'Borealis' Branches horizontal, less crowded; leaves longer and narrower.

'Expansa' Growth dwarf; branches spreading along the ground, with tips turned up.

'Fennica' Foliage dark green.

'Frohnleiten' Growth dwarf, irregular; leaves short.

'Gnom' Growth dwarf, conical, wide, dense; branches and leaves crowded.

'Microphylla'

'Minima' Growth dwarf.

'Nana' Growth dwarf, wider than high when young, later conical, dense; branches horizontal.

'Pendula' Branches drooping, even at the ends.

'Pimoko' Growth dwarf, irregular.

11. *P. orientalis* (L.) Link.   Oriental spruce.

A beautiful, slow-growing, short-leaved native of Asia that is successful when planted in northeastern North America. Foliage dark green and dense. Leaves 7–12 mm (¼–⅖ in) long. Cones 6–9 cm (2½–3½ in) long.

'Atrovirens' Leaves shining dark green.

'Aurea' Leaves on young shoots yellow, changing to green.

'Aurea Compacta' Growth dwarf, spreading to taller and narrow; leaves yellow on upper surface.

'Aureospicata' New growth yellow-green, then green.

'Bergman's Gem'

'Bergman's Repens' Similar to 'Nana Compacta'.

'Compacta' Growth conical, compact; foliage tending to be brown.

'Compacta Aurea' Same as 'Aurea Compacta'.

'Early Gold' Growth dwarf, spreading; foliage yellow in spring.

'Gowdy' Growth slow, columnar; leaves small.

'Gracilis' Growth dwarf, columnar.

'Inversa'

'Martin'

'Mount Vernon' Growth dwarf, rounded.

'Nana' Growth dwarf, rounded, dense, irregular.

'Nana Compacta' Growth slow, more rapid than 'Nana'; rounded or flat-topped or conical.

'Nigra Compacta' Growth more rapid than 'Nana Compacta'.

'Nutans' Shrub; growth conical.

'Pendula' Growth compact; branches drooping.

'Raraflora Fuke'

'Repens' Growth slow.

'Skylands' Same as 'Aurea Compacta'.

'Weeping Dwarf' Growth dwarf; branches drooping.

*12. *P. pungens* Engelm. Colorado blue spruce. Blue spruce.

A native of the Rocky Mountains, probably the most popular and abundantly cultivated of the spruces, next to Norway spruce. Easily recognized by its distinctive strong, stiff branches with sharp-pointed, stiff, blue leaves. Successful for many ornamental purposes but slow growing. Leaves 2–4 cm (¾–1½ in) long. Cones 6–10 cm (2½–4 in) long.

'Argentea' Leaves lighter blue.

'Aurea' Leaves yellow.

'Bakeri' Leaves longer.

'Bismarck'

'Blue Bun'

'Blue Spreader'   Growth low, spreading, with occasional vigorous, upright branches.

'Blue Trinket'   Growth dwarf, conical, compact.

'Columnaris'   Growth columnar; branches very short.

'Compacta'   Shrub; growth dense, slow, flat-topped.

'Compact Thurl'

'Egyptian Pyramid'   Growth dwarf, dense.

'Elegantissima'   Young leaves white.

'Endtz'   Growth conical, dense; twigs short, upright.

'Eric Frahm'   Growth conical, regular.

'Fat Albert'   Growth dense.

'Flavescens'   Leaves white to pale yellow.

'Formidable'

'Fox Tail'   New shoots with short and parallel leaves at the tips of the shoots becoming long and perpendicular at the base of the shoots.

'Furst Bismarck'   Foliage blue.

'Glauca'   Leaves blue.

'Glauca Compacta'   Growth dwarf, conical, dense.

'Glauca Globosa'   Growth dwarf, rounded.

'Glauca Hoopsii'   Same as 'Hoopsii'.

'Glauca Pendula'   Branches and twigs drooping, the tips turning up.

'Glauca Procumbens'   Growth dwarf; branch tips turning up.

'Glauca Prostrata'   Growth dwarf, spreading over the ground.

'Globosa'   Growth dwarf, rounded, dense.

'Goldie'   Same as 'Walnut Glen'.

'Gotelli's Broom'   Growth dwarf, flat-topped.

'Hoopsii'   Growth rapid, conical, dense; leaves very blue.

'Hoto'   Growth conical, dense.

'Hunnewelliana'   'Growth dwarf, conical, dense; branches erect.

'Iseli Fastigiate'   Growth columnar.

'Iseli Fat Albert'   Same as 'Fat Albert'.

'Iseli Foxtail'   Same as 'Fox Tail'.

'Kosteri'   Growth conical; twigs drooping; leaves blue.

'Kosteri Pendula'   Branches drooping; leaves blue.

'Lombarts'   Foliage blue.

'Lutea'   Foliage yellow.

'Luusbarg'   Growth dwarf, wide spreading.

'Mission Blue'   Growth conical, dense.

'Moerheimii'   Growth compact; leaves blue.

'Moll'   Growth dwarf, conical, compact, as wide as high.

'Montgomery'   Growth dwarf, rounded becoming conical, compact.

'Mrs. Cessarini'   Growth dwarf; leaves small.

'Nana'   Growth dwarf.

'Oldenburg'   Growth conical, regular; branches yellow-brown.

'Pendula'   Branches drooping.

'Procumbens'   Growth low, spreading.

'Prostrata'   Growth low, spreading.

'Prostrate Blue Mist'   Growth dwarf, low, spreading.

'Royal Knight'   Growth conical, compact.

'St. Mary'   Same as 'St. Mary's Broom'.

'St. Mary's Broom'   Growth dwarf, conical, wider than high.

'Schovenhorst'   Growth conical; branches many, crowded.

'Schwartz'   Terminal buds, prominent, thick, yellow-brown.

'Select'

'Special'

'Spek'   Leaves blue-white.

'Thompsonii'   Same as 'Thomsen'.

'Thomsen'   Growth conical; leaves long, silver-blue.

'Thume'   Growth conical; leaves curved forward; similar to 'Montgomery' but growth more rapid.

'Viridis'   Leaves dull green.

'Vuyk'   Buds thick, long, bright brown.

'Walnut Glen'   Foliage glaucous, patched with white in spring.

'Yvette'

'Ziegler'

Picea pungens Cultivar Character Grouping

Dwarf

'Blue Trinket'	'Hunnewelliana'	'Nana'
'Egyptian Pyramid'	'Moll'	'Prostrate Blue Mist'
'Globosa'	'Montgomery'	'St. Mary'
'Gotelli's Broom'	'Mrs. Cessarini'	'St. Mary's Broom'

+13.   *P. rubens* Sarg.   Red spruce.

An abundant native of northeastern North America that is rarely cultivated. Does not endure dry conditions well. Leaves crowded, incurved on the branch. Outer bud scales with very long awl-shaped points. Buds generally concealed by the foliage. Leaves 13 mm (½ in) long. Cones 3–5 cm (1¼–2 in) long.

'Monstrosa'   Twigs partially fused together.
'Nana'   Growth dwarf.
'Pocono'   Growth compact; foliage tufted.
'Virgata'   Branches long, slender, lacking twigs.

14.   *P. sitchensis* Carr.   Sitka spruce.   Tideland spruce.

A timber tree native along the coast of northwestern North America. Hardy in the Northeast but not easy to grow, usually requiring irrigation to survive the hot, dry conditions of summer. Leaves 1.5– 2.5 cm (½–1 in) long. Cones 6–10 cm (2½–4 in) long.

'Aurea'   Leaves yellow on upper surface.
'Compacta'   Growth dwarf, compact.
'Papoose'   Growth dwarf, conical.
'Strypemonde'   Growth dwarf, rounded.
'Tenas'   Growth dwarf; similar to 'Papoose'.
'Upright Dwarf'   Growth dwarf, conical, compact.

15.   *P. torano* Koehne.   Tigertail spruce.

Native to Japan. Rarely planted tree with thick, rigid branches and leaves. Leaves higher than wide, with spiny tips, 1.3–2.5 cm (½–1 in) long. Cones 10–13 cm (4–5 in) long.

Pinus strobus

Pinus L.　Pine.

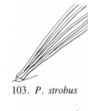

103. *P. strobus*

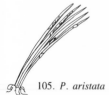

104. *P. flexilis*

105. *P. aristata*

The pines, the largest genus of conifers, include about 100 species, nearly all of which are native north of the equator. They are abundantly planted for many purposes (as Christmas trees, ornamentals, hedges, and timber and for resin). These trees often grow well on poor or thin soils and under better conditions can be excellent, fast-growing trees. The pines are distinguished by their bundles of two, three, four, or five leaves (one species has solitary leaves) and by the dry paper sheaths surrounding the base of the leaf bundles. The female cones are a prominent feature. Identification without cones is difficult but is perhaps possible in the case of the species covered here.

1. Leaves in bundles of 5 (Fig. 103).
　　2. Edges of leaves smooth and unbroken (Fig. 104); rare in cultivation.
　　　　3. Leaves 2.5–4.0 cm (1–1½ in) long, usually with white, sticky resin spots (Fig. 105); sheaths at base of leaf bundles remaining on the plant for 2–3 years . . . . *P. aristata*
　　　　　　　　　　　　　　　　　　　　　　　Bristlecone pine

118

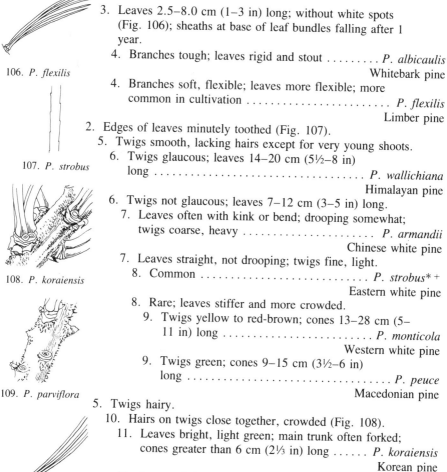

106. *P. flexilis*

107. *P. strobus*

108. *P. koraiensis*

109. *P. parviflora*

110. *P. ponderosa*

111. *P. nigra*

112. *P. bungeana*

3. Leaves 2.5–8.0 cm (1–3 in) long; without white spots (Fig. 106); sheaths at base of leaf bundles falling after 1 year.

    4. Branches tough; leaves rigid and stout . . . . *P. albicaulis*
                                                  Whitebark pine

    4. Branches soft, flexible; leaves more flexible; more common in cultivation . . . . *P. flexilis*
                                                  Limber pine

2. Edges of leaves minutely toothed (Fig. 107).

   5. Twigs smooth, lacking hairs except for very young shoots.

     6. Twigs glaucous; leaves 14–20 cm (5½–8 in) long . . . . *P. wallichiana*
                                      Himalayan pine

     6. Twigs not glaucous; leaves 7–12 cm (3–5 in) long.

       7. Leaves often with kink or bend; drooping somewhat; twigs coarse, heavy . . . . *P. armandii*
                                   Chinese white pine

       7. Leaves straight, not drooping; twigs fine, light.

        8. Common . . . . *P. strobus** +
                                   Eastern white pine

        8. Rare; leaves stiffer and more crowded.

         9. Twigs yellow to red-brown; cones 13–28 cm (5–11 in) long . . . . *P. monticola*
                                 Western white pine

         9. Twigs green; cones 9–15 cm (3½–6 in) long . . . . *P. peuce*
                                   Macedonian pine

   5. Twigs hairy.

     10. Hairs on twigs close together, crowded (Fig. 108).

       11. Leaves bright, light green; main trunk often forked; cones greater than 6 cm (2⅓ in) long . . . . . . *P. koraiensis*
                                     Korean pine

       11. Leaves dark green; cones less than 5 cm (2 in) long.

         12. Leaves 4.5–8.0 cm (1¾–3⅛ in) long; some main branches usually spreading along ground . . . . *P. pumila*
                                   Dwarf stone pine

         12. Leaves 5–13 cm (2–5 in) long; branches horizontal, but not spreading along ground . . . . *P. cembra*
                                   Swiss stone pine

     10. Hairs on twigs sparse (Fig. 109) . . . . *P. parviflora*
                                  Japanese white pine

1. Leaves in bundles of 2 or 3 (rarely 4) (Fig. 110 and 111).

  13. Sheaths at the base of the leaf bundles absent (Fig. 112).

113. *P. sylvestris*

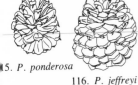

114. *P. rigida*

15. *P. ponderosa*

116. *P. jeffreyi*

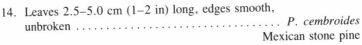

117. *P. virginiana*

118. *P. banksiana*

119. *P. sylvestris*

120. *P. echinata*

121. *P. sylvestris*

14. Leaves 2.5–5.0 cm (1–2 in) long, edges smooth, unbroken . . . . *P. cembroides*
Mexican stone pine

14. Leaves 5–10 cm (2–4 in) long, edges toothed; bark flaky, with white patches . . . . *P. bungeana*
Lacebark pine

13. Sheaths at base of leaf bundles present (Fig. 113).

15. Leaves in bundles of 3 (Fig. 114).

16. Leaves 12–27 cm (4¾–11 in) long.

17. Twigs not glaucous; cone scales with prickles turned outward when the cone is dry (Fig. 115); cones 8–15 cm (3–6 in) long . . . . *P. ponderosa*
Ponderosa pine

17. Twigs glaucous; cone scales with prickles not projected beyond the outline of the dry cone (Fig. 116), cones 13–35 cm (5–12 in) long . . . . *P. jeffreyi*
Jeffrey pine

16. Leaves 7–14 cm (3–5½ in) long.

18. Leaves flexible, glaucous, often 2 to a bundle.

19. Young twigs glaucous; leaves straight . . . . . *P. echinata* [+]
Shortleaf pine

19. Young twigs not glaucous; leaves twisted . . . . *P. tabulaeformis*
Chinese pine

18. Leaves stiff, not glaucous, always 3 in a bundle . . . . *P. rigida* [+]
Pitch pine

15. Leaves in bundles of 2 (Fig. 117).

20. Leaves mostly less than 7.5 cm (3 in) long.

21. Leaves 2.5 cm (1 in) long; leaf sheaths 2 mm (¹⁄₁₆ in) long (Fig. 118) . . . . *P. banksiana* [+]
Jack pine

21. Leaves mostly greater than 2.5 cm (1 in) long; leaf sheaths usually greater than 2 mm (¹⁄₁₆ in) long (Fig. 119).

22. Sheaths at base of leaf bundles mostly 2–4 mm (¹⁄₁₆– ⅛ in) long; leaves light green.

23. Leaves straight, not strongly twisted (Fig. 120) . . . . *P. echinata* [+]
Shortleaf pine

23. Leaves usually twisted (Fig. 121).

24. Leaves coarse, stiff, in bundles of 3 and 2.

25. Leaves frequently greater than 7.5 cm (3 in) long . . . . *P. tabulaeformis*
Chinese pine

25. Leaves usually less than 7.5 cm (3 in) long . . . . *P. pungens*
Table-mountain pine

122. *P. sylvestris*

123. *P. contorta*

124. *P. leucodermis*

125. *P. mugo*

126. *P. resinosa*

127. *P. nigra*

128. *P. nigra*

129. *P. thunbergiana*

24. Leaves fine, usually thinner, in bundles of 2 only.
  26. Young twigs glaucous . . . . *P. virginiana*[+]
                                       Virginia pine
  26. Young twigs not glaucous.
    27. Common; bark orange-red; bud scales free, often fringed at the tips (Fig. 122); leaves often bluish . . . . *P. sylvestris**
                                       Scotch pine
    27. Rare; bark brown or darker; bud scales pressed tightly together (Fig. 123); leaves light green . . . . *P. contorta*
                                    Lodgepole pine
22. Sheaths at base of leaf bundles mostly more than 5 mm (⅛ in) long; leaves dark green.
  28. Trees; single-stemmed (Fig. 124), slow growing; leaves very dark green; bud scale tips pale . . . . *P. leucodermis*
                                    Bosnian pine
  28. Shrubs; multistemmed or many-branched (Fig. 125); leaves green; bud scale tips brown . . . . *P. mugo**
                                     Mugo pine
20. Leaves mostly more than 7.5 cm (3 in) long.
  29. Leaves very stiff, more than 2 mm (1/16 in) wide . . . . *P. pinaster*
                                    Cluster pine
  29. Leaves stiff, 1 mm (1/32 in) wide.
    30. Leaves in bundles of 3 and 2; rare.
      31. Young twigs glaucous; leaves straight . . . . *P. echinata*[+]
                                  Shortleaf pine
      31. Young twigs not glaucous; leaves twisted . . . . *P. tabulaeformis*
                                  Chinese pine
    30. Leaves in bundles of 2 only; common.
      32. Fresh leaves snap or break when bent (Fig. 126) . . . . *P. resinosa*[+]
                                  Red pine
      32. Fresh leaves merely fold when bent (Fig. 127).
        33. Leaves straight, not twisted, at least 7.5 cm (3 in) long; bud scales pressed tightly together.
          34. Leaves stiff, coarse, dark green; young twigs not glaucous; bark black.
            35. Buds light brown, resinous; growth uniform (Fig. 128); leaves very dark green . . . . *P. nigra**
                                  Austrian pine
            35. Buds white, not resinous; trunk and branches often contorted or forked (Fig. 129); leaves a lighter green . . . . *P. thunbergiana*
                              Japanese black pine

130. *P. leucodermis*

131. *P. mugo*

34. Leaves somewhat flexible, thinner, light green; young
   twigs slender, glaucous; bark orange-red . . . . *P. densiflora*
   Japanese red pine
33. Leaves usually twisted, frequently less than 7.5 cm (3 in) long.
   36. Leaves very dark green; usually a shrub.
      37. Plant single-stemmed (Fig. 130); bud scale tips
         pale . . . . *P. leucodermis*
         Bosnian pine
      37. Plant many-stemmed (Fig. 131), spreading, bud scale
         tips brown . . . . *P. mugo**
         Mugo pine
   36. Leaves light green, often bluish; usually a
      tree . . . . *P. sylvestris**
      Scotch pine

1. *P. albicaulis* Engelm.   Whitebark pine.

Native to the high altitudes of western United States. Very rarely
cultivated. Leaves and branches very similar to those of *P. flexilis*.
Branches tougher and cones much shorter than *P. flexilis*. Leaves 5
to a bundle, 4.0–6.5 cm (1½–2½ in) long. Cones nearly round 4–7
cm (1½–2¾ in) long.

'Flinck'   Growth dwarf.
'Nana'   Growth dwarf.
'Nobles Dwarf'   Growth dwarf, compact.
'Number One Dwarf'   Growth dwarf.

2. *P. aristata* Engelm.   Bristlecone pine.

A low and shrubby native of western United States, difficult to
cultivate. Leaves 5 to a bundle, with entire edges, 2.5–4.0 cm (1–
1½ in) long, often with scattered white spots of resin. Cones 8–9 cm
(3–3½ in) long with spines.

'Baldwin Dwarf'   Same as 'Cecilia'.
'Cecilia'   Growth dwarf, congested, rounded.
'Sherwood Compact'   Growth dwarf, rounded; leaves short.

3. *P. armandii* Franch.   Chinese white pine.   Armand pine.

Native to eastern Asia, cultivated mostly only in arboreta. Leaves 5
to a bundle, 8–15 cm (3–6 in) long, drooping or spreading, some-

times with a kink or bend near the base, with minutely toothed edges, lacking white bands on the back. Cones 10–18 cm (4–8 in) long, with thick scales.

+4.   *P. banksiana* Lamb.   Jack pine.

Native to Canada and northeastern United States. A scraggly pine of little use for ornamental planting or Christmas trees. Leaves 2 to a bundle, twisted, stiff, 2.5 cm (1 in) long. Cones 4–5 cm (1½–2 in) long, lacking prickles.

'Chippewa'   Growth irregular, rounded; branches stiff, spreading.
'Fastigata'
'Manomet'   Growth rounded, low; leaves small, may be yellow.
'Nana'
'Neponset'   Growth low, flat-topped.
'Schoodic'   Growth low, spreading over ground.
'Uncle Fogy'   Growth dwarf, spreading, close to ground; leaves
   larger.
'Watt's Golden'
'Wisconsin'   Shrub; growth rounded, dense.

5.   *P. bungeana* Zucc. ex Endl.   Lacebark pine.

Native to northwestern China. Slow growing but can make an interesting ornamental because of its shiny green leaves, uniform growth when young, and gray green-mottled, flaking bark. Leaves 3 to a bundle, 5–10 cm (2–4 in) long. Cones lacking stalks, 5–8 cm (2–3 in) long.

6.   *P. cembra* L.   Swiss stone pine.

Native to Soviet Union and the Alps of Europe. Hardy and slow growing. Makes an excellent ornamental because of its dense dark green foliage and symmetrical growth. Leaves 5 to a bundle, 5–13 cm (2–5 in) long. Cones short stalked, 5–9 cm (2–3½ in) long. Twigs densely hairy.

'Aurea'   Foliage yellow in winter.

'Aureovariegata'   Leaves more or less yellow.

'Blue Mound'   Growth slow.

'Broom'

'Chalet'   Growth dwarf, columnar, dense.

'Chamolet'   Growth slow, dense.

'Chlorocarpa'   Growth slow, dense, erect, bushy.

'Columnaris'   Growth columnar.

'Compacta'   Growth slow.

'Compacta Glauca'   Growth slow, conical, compact; leaves glaucous; probably same as 'Compacta'.

'Glauca'   Leaves glaucous.

'Globe'   Growth dwarf, rounded, compact.

'Jermyns'   Growth dwarf, conical, compact.

'Kairamo'   Leaves close together and sharply hanging down at ends of branches.

'Prostrata'   Growth spreading.

'Pygmaea'   Growth slow, irregular, flat-topped; leaves short, irregular in length.

'Silver Sheen'   Foliage blue.

'Stricta'   Growth columnar.

7. *P. cembroides* Zucc.   Mexican stone pine.

Small tree native to southwestern United States and Mexico. Not often cultivated. Leaves in bundles of 3 or sometimes 2, 2–5 cm (1–2 in) long. Cones 2.5–5.0 cm (1–2 in) long. Seeds narrowly winged, 2 cm (¾ in) long or less.

8. *P. contorta* Dougl. ex Loud.   Lodgepole pine.   Shore pine.   Western scrub pine.

Native to the western United States. Four botanical varieties, none of which is frequently planted. Leaves 2 to a bundle, 3–9 cm (1–3½ in) long. Cones 2.5–5.0 cm (1–2 in) long.

'Compacta'   Growth dense, compact.

'Frisian Gold'   Leaves yellow.

'Goldchen' Foliage yellow.
'Minima' Same as 'Spaan's Dwarf'.
'Spaan's Dwarf' Growth dwarf, as wide as high.

9. *P. densiflora* Sieb. & Zucc. Japanese red pine.

Native to Japan. A hardy tree, known more as a large shrub in cultivation in northeastern North America. Leaves 2 to a bundle, 8–13 cm (3–5 in) long. Cones 5 cm (2 in) long. Seeds 7 mm (¼ in) long.

'Alice Verkade' Growth dwarf, spreading, bun-shaped, dense.
'Aurea' Leaves on some branches yellow in fall and winter.
'Globosa' Shrub; growth rounded, dense.
'Griffith Prostrate' Similar to 'Pendula'.
'Heavy Bud' Growth dwarf, rounded; buds larger.
'Hospitalis' Leaves shorter.
'Jane Kluis' Growth dwarf, rounded.
'Oculus-draconis' Leaves with alternate broad, horizontal yellow rings.
'Pendula' Growth spreading over the ground; branches twisted, drooping, brittle.
'Tanyosho' Growth rounded, irregular.
'Tanyosho Special' Growth dense.
'Umbraculifera' Growth dwarf, weak; branches crowded.

+10. *P. echinata* Mill. Shortleaf pine.

Native only as far north as southeastern New York. Generally not cultivated in the Northeast. Leaves 2, sometimes 3 to a bundle, 8–13 cm (3–5 in) long. Cones 4–6 cm (1½–2½ in) long. Seeds 5–7 mm (⅕–¼ in) long.

'Clines Dwarf' Growth dwarf, covering ground.

11. *P. flexilis* James. Limber pine.

Native to the mountains of western United States. A flexible, hardy species that can grow rapidly and has a loose, open character.

Branches often long, somewhat twisted, and sparsely foliated. Leaves 5 to a bundle, 4–8 cm (1½–3 in) long. Cone 8–15 cm (3–6 in) long, sometimes as long as 25 cm (10 in).

'Albovariegata'   Foliage with patches of white.
'Bergman Dwarf'   Growth slow.
'Compacta'   Growth slow, conical.
'Extra Blue'   Foliage very blue.
'Glauca'   Leaves glaucous.
'Glauca Pendula'   Same as 'Pendula'.
'Glenmore'   Same as 'Glenmore Dwarf'.
'Glenmore Dwarf'   Growth slow, conical, dense; leaves coarser.
'Nana'   Growth dwarf; leaves short.
'Pendula'   Growth rapid, spreading; main trunk and branches drooping.
'Pendula Glauca'   Same as 'Pendula'.
'Reflexa'   Leaves longer.
'Temple'   Growth slow, conical.
'Tiny Temple'   Growth low, spreading.
'Vanderwolf's Pyramid'   Growth rapid, conical.
'Witch's Broom'   Growth dwarf, compact, dense.

12. *P. jeffreyi* Grev. & Balf.   Jeffrey pine.

A native of Oregon and California that makes a symmetrical, long-leaved, coarse ornamental. Leaves 3 to a bundle, 13–20 cm (5–8 in) long. Cones 13–35 cm (5–12 in) long. Seeds 13 mm (½ in) long.

13. *P. koraiensis* Sieb. & Zucc.   Korean nut pine.   Korean pine.

Native to Korea and Japan. A hardy pine with slow growth that may fork. Leaves 5 (sometimes 3) to a bundle, dense, dark green, with blue tint on the inner sides, 6–10 cm (2½–4 in) long. Cones with thick scales falling from tree before opening, 10–15 cm (4–6 in) long.

'Avogadro'   Probably same as 'Tabuliformis'.
'Compacta Glauca'   Branches short, thick; leaves glaucous.
'Dwarf'   Growth slow.

'Glauca'   Leaves thick, strongly glaucous.

'Jack Corbit'

'Nana'

'Oculus-draconis'   New foliage with yellow rings.

'Tabuliformis'   Growth flat-topped.

'Tortuosa'   Leaves spirally twisted.

'Winton'   Growth dwarf, wider than high; winter buds larger.

14. *P. leucodermis* Antoine.   Bosnian pine.   Graybark pine.

Native to the Balkans. Not commonly cultivated. Slow growing but erect and single-stemmed. Leaves 2 to a bundle, very dark green, 4.0–6.5 cm (1½–2½ in) long. Cones 7.5 cm (3 in) long.

'Aureospicata'   Leaves yellow at tips in spring.

'Compact Gem'   Growth dwarf, compact; leaves curved.

'Pygmy'   Similar to 'Schmidtii'.

'Satellit'   Growth conical.

'Schmidtii'   Growth dwarf, conical to rounded, dense.

15. *P. monticola* Dougl.   Western white pine.

Native to western North America. Rare in cultivation. Difficult to distinguish from *P. strobus*. Leaves stiffer and growth more dense. Leaves 5 to a bundle, 4–10 cm (1½–4 in) long. Cones 10–25 cm (4–10 in) long.

'Ammerland'   Growth rapid; branches thick.

'Glauca'   Leaves glaucous.

'Minima'   Growth compact; leaves short, crowded, dark green.

'Pendula'   Branches drooping.

'Pygmaea'

*16. *P. mugo* Turra.   Mugo pine.   Swiss mountain pine.

A native European shrub planted frequently as a dwarf conifer. Hardy and slow growing, with crowded, twisted, dark green leaves.

Leaves 2 to a bundle, 2–5 cm (¾–2 in) long. Cones 2.5–4.0 cm (1–1½ in) long.

'Allen's Seedling'   Growth dwarf.

'Alpenglow'   Growth dwarf, vigorous, compact.

'Aurea'   Growth dwarf; leaves longer, turning yellow in winter.

'Brevifolia'   Same as 'Kissen'.

'Compacta'   Growth dwarf, compact.

'Corley's Mat'   Growth low, flat-topped.

'Elfengren'   Growth dwarf, compact.

'Emerald Tower'   Growth slow, upright.

'Frisia'   Growth slow; branches upright, crowded.

'Gnom'   Growth dwarf; twigs many.

'Green Candle'   Growth dwarf, columnar.

'Hesse'   Growth dwarf, rounded, dense.

'Humpy'   Growth dwarf, compact; leaves short.

'Iseli White Bud'   Winter buds white.

'Kissen'   Growth dwarf, bun-shaped, dense.

'Knapenburg'   Growth dwarf, irregular, dense.

'Kobold'   Growth dwarf, rounded, dense.

'Kokarde'   Leaves with alternating yellow and green rings.

'Mayfair Dwarf'

'Mops'   Growth dwarf, rounded.

'Ophir'   Growth slow, rounded; upper surface of leaf tips yellow.

'Oregon Jade'   Growth dwarf, compact.

'Oregon Pixie'   Growth dwarf, compact.

'Pal Maleter'

'Prostrata'

'Pyramidalis'

'Rigi'   Growth columnar.

'Sherwood Compact'   Growth dwarf.

'Slavinii'   Growth dwarf; branches erect.

'Spaan'   Growth low, spreading; leaves very short, appearing to have cutworm damage.

'Spaan's Pygmy'   Same as 'Spaan'.

'Teeny'   Growth dwarf, mounded, compact.

'Trompenburg'   Growth dwarf.

'Tyrol'   Growth dwarf, upright, open.

'Valley Cashion'   Growth low; leaves short.

'Winter Gold'   Growth dwarf, open; leaves bright yellow in winter, twisted.

*17.   P. nigra Turra.   Austrian pine.

An eastern and southern European pine. Planted often as an ornamental, in hedges, and commercially for Christmas trees. A coarse tree with dark green leaves. Leaves 2 to a bundle, 7.7–16.5 cm (3–6½ in) long. Cones 5–10 cm (2–4 in) long. Seeds 7 mm (¼ in) long.

'Aurea'   Young leaves yellow.

'Aureovariegata'   Leaves variegated yellow.

'Balcanica'   Growth dwarf, cushion-shaped; leaves thin, crowded.

'Bujotii'   Growth dwarf, conical; twigs short.

'Columnaris'   Growth columnar; branches short.

'Compacta'   Growth compact.

'Contorta'   Shrub; leaves twisted.

'Globosa'   Growth rounded, strong; branches short.

'Hornibrookiana'   Shrub; growth slow, wider than high; leaves crowded, short.

'Jeddeloh'   Growth compact; branches and twigs short.

'Monopeliensis'

'Monstrosa'   Branches short, contorted; leaves crowded.

'Nana'   Growth dwarf, dense, as wide as high.

'Pendula'   Branches drooping.

'Porretiana'

'Prostrata'   Growth low; branches close to the ground.

'Pygmaea'   Growth slow, rounded, dense.

'Pyramidalis'   Branches ascending.

'Schovenhorst'   Similar to 'Strypemonde'.

'Strypemonde'   Growth coarse, open.

'Tenuifolia'

'Trompenburg'

18.   P. parviflora Sieb. & Zucc.   Japanese white pine.

Native to Japan and Taiwan. Increasingly popular for dwarf and slow-growing cultivars. Leaves 5 to a bundle, usually twisted, crowded in tufts at twig ends, 2–4 cm (¾–1½ in) long. Cones often produced early, 5–8 cm (2–3 in) long. Seeds less than 1.3 cm (½ in) long.

'Adcock's Dwarf'   Growth dwarf, dense, higher than wide; leaves short.

'Aizu'   Growth slow; branches and leaves thick.

'Ara-kawa'   Bark rough; used for bonsai grafts.

'Baldwin'   Growth dwarf; leaves blue.

'Bergmanii'   Growth dwarf, rounded, wide, spreading; leaves blue.

'Blue Giant'

'Brevifolia'   Growth narrow, ascending; branches and twigs few; leaves short.

'Compacta'   Growth slow.

'Emperor'   Foliage emerald green.

'Fukuzumi'   Growth slow, rounded.

'Gimborn's Ideal'   Shrub; branches ascending; foliage dense, glaucous.

'Gimborn's Pyramid'   Growth slow, conical; foliage blue.

'Gi-on'   Growth slow, compact; leaves thick, straight.

'Glauca'   Growth slow, upright, narrow; branches on young plants few; leaves stiffer, twisted, blue.

'Glauca Compacta'   Growth dwarf.

'Glauca Nana'   Growth dwarf.

'Ha-zumari-goyo'   Similar to 'Brevifolia'.

'Ibo-can'   Bark with wartlike pattern.

'Janome-ibokan-goyo'   Bark rough, warty; leaves blue.

'Joe Burke'

'Kokonoe'   Growth dwarf; leaves thick, twisted.

'Ko-raku'   Growth dwarf, compact; leaves yellow-green.

'Momo-yama'   Leaves thick.

'Nana'   Growth dwarf; leaves short.

'Nasu'   Growth dwarf.

'Negishi'   Cones produced early.

'Ogon-janome'   Leaves with yellow horizontal rings.

'Saki-shiro-goyo'   Leaf tips white.

'Setsu-gek-ka'   Growth dwarf; leaves short, thick, dark green.

'Shiko-ku-goyo'   Bark rough; leaves short.

'Shiobara-goyo'   Growth dwarf; leaves thin, short, light green.

'Shi-on'   Growth rapid; leaves yellow-green, twisted.

'Tempelhof'   Growth rapid; foliage glaucous.

'Venus'   Growth dwarf.

'Wells'   Foliage glaucous.
'Yu-ho'   Growth dwarf, spreading over the ground.
'Zui-sho'   Growth dwarf; leaves thin, short.

19.   *P. peuce* Griseb.   Macedonian pine.   Macedonian white pine.

Native to the Balkans. Grown successfully in the Northeast. Foliage similar to that of *P. strobus*, but habit of tree tends to be narrow and ascending, as in *P. cembra*. Leaves in bundles of 5, 7.7–10.0 cm (3–4 in) long. Cones 9.0–15.5 cm (3½–6 in) long.

'Arnold Dwarf'   Growth dwarf, higher than wide.
'Glauca Compact'   Growth compact; leaves glaucous.
'Nana'   Same as 'Arnold Dwarf'.

20.   *P. pinaster* Ait.   Cluster pine.   Maritime pine.

Native to the Mediterranean area. Rare in cultivation in the Northeast but widely planted in other parts of the world. Leaves in bundles of 2, 12–23 cm (5–9 in) long, more than 2 mm (1/16 in) wide. Cones 10–18 cm (4–7 in) long.

21.   *P. ponderosa* Dougl. ex P. Laws & C. Laws.   Ponderosa pine.   Yellow pine.

Native to western North America. Cultivated infrequently. Branches coarse, stout, heavy. Leaves 3 in a bundle, 13–28 cm (5–11 in) long. Cones 8–15 cm (3–6 in) long. Seeds 7 mm (¼ in) long.

'Canyon Ferry'   Growth dwarf.
'Pendula'   Branches contorted.
'Tortuosa'   Branches contorted.

22.   *P. pumila* Regel.   Dwarf stone pine.

Native to northeastern Siberia, Korea, and Japan. Once considered a variety of *P. cembra*. Shrub that can be grown in the Northeast with some success. Leaves 4.5–8.0 cm (1¾–3⅛ in) long. Cones 3.3–4.5 cm (1¼–1¾ in) long.

'Dwarf Blue'   Growth dwarf.
'Glauca'   Growth slow, rounded; leaves gray.
'Hillside'   Growth slow.
'Prostrata'   Growth spreading over the ground.

23.   *P. pungens* Lamb.   Table-mountain pine.

Native to the southeastern United States, a small tree infrequently cultivated. Leaves 2 to a bundle, twisted, dark green 3–6 cm (1¼–2½ in) long. Cones 6–9 cm (2½–3½ in) long, usually three together. Seeds 7 mm (¼ in) long.

+24.   *P. resinosa* Ait.   Red pine.   Norway pine.

Native to northeastern North America. Planted for timber, as a windbreak, and occasionally as an ornamental. Leaves 2 to a bundle, 7.5–15.0 cm (3–6 in) long. Cones 4–6 cm (1½–2½ in) long. Seeds 3 mm (⅛ in) long.

'Aurea'   Foliage yellow.
'Don Smith'   Growth dwarf, low, wider than high; branches spreading, then upright; cones produced early.
'Globosa'   Growth dwarf; branches and leaves crowded.
'Nobska'   Growth dwarf.
'Quinobeguin'   Growth slow, rounded, low.

+25.   *P. rigida* Mill.   Pitch pine.

Native to northeastern North America. Infrequently cultivated because of its scraggly appearance. Leaves 3 to a bundle, 5–13 cm (2–5 in) long. Cones 5–10 cm (2–4 in) long. Seeds 7 mm (¼ in) long.

'Aurea'   Leaves yellow.
'Hillside Weeping'   Branches drooping.
'Sherman Eddy'   Growth conical, compact.

* + 26. *P. strobus* L. Eastern white pine.

Native to northeastern North America, once one of the region's most prominent conifirs. A fast-growing hardy tree that can be used for nearly all planting purposes. Leaves 5 to a bundle, soft with bluish or grayish tint, 5–13 cm (2–5 in) long. Cones 8–20 cm (3–8 in) long.

'Alba' Leaves erect, pale.

'Amelia's Dwarf' Branches slightly drooping.

'Anna Feile' Growth slow, rounded.

'Aurea' Leaves yellow on young twigs.

'Bennett's Clump Leaf' Leaves of each bundle closed, the bundle thus appearing to be a single leaf.

'Bennett's Contorted' Growth dwarf, higher than wide; branches drooping.

'Bennett's Dragon Eye' Leaves yellow, then green, with yellow horizontal rings.

'Bergman's Variegated' Leaves with horizontal yellow rings.

'Bloomer's Dark Globe' Same as 'Bloomer's Globe'.

'Bloomer's Globe' Growth slow, open; branches ascending; foliage dark green.

'Blue Mist'

'Blue Shag' Growth dwarf, wide, dense; leaves short.

'Brevifolia' Shrub; growth dwarf, flat-topped.

'Compacta' Growth rounded, becoming conical, compact.

'Contorta' Leaves twisted.

'Contorta Nana' Growth dwarf; branches irregular.

'Crazy Form' Growth twisted, contorted.

'Curtis' Same as 'Curtis Dwarf'.

'Curtis Dwarf' Growth spreading, compact, dense; leaves short.

'Densa' Growth dwarf, dense, compact; similar to 'Brevifolia'.

'Dove's Dwarf' Growth dwarf, wide.

'Elf' Growth slow; leaves glaucous.

'Elkin's Dwarf'

'Fastigiata' Growth columnar, becoming conical.

'Glauca' Leaves glaucous.

'Globosa' Growth rounded.

'Gracilis Viridis' Leaves thinner, lighter green.

'Green Shadow'   Growth dwarf, rounded; foliage dense.

'Greg's Form'   Growth slow, rounded.

'Helen'

'Hershey'   Growth compact, irregular.

'Hillside Gem'   Growth dwarf, higher than wide, open; leaves short.

'Hillside Weeper'   Growth irregular, trunk twisted; branches irregular; rare, does not graft.

'Hillside Winter Gold'   Foliage yellow in winter.

'Horsford'   Growth dwarf, rounded; leaves thin, long.

'Horsham'   Growth slow; cones produced at an early age.

'Inversa'   Branches drooping.

'Jericho'   Shrub; branches thin.

'Julian's Dwarf'   Growth slow, conical.

'Lenmore'   Growth rapid, conical.

'Lone Pine Broom'

'Macopin'   Growth strong, dense, conical to rounded; twigs thick; leaves long.

'Martin's Broom'

'Merrimack'   Growth low, dense; leaves short.

'Minima'   Growth dwarf, mound-shaped.

'Minuta'   Growth dwarf, open.

'Monophylla'   Leaves joined to appear as one.

'Nana'   Growth usually dwarf, rounded or conical, as wide to twice as wide as high.

'Northway Broom'   Growth dwarf, flat-topped; leaves narrow.

'Ontario'   Growth dwarf or normal rate, depending on whether grafted from rapid growth or from slow growth, rounded; leaves short.

'Pendula'   Growth rapid; branches drooping and long.

'Prostrata'   Growth very low; branches spreading over the ground.

'Pygmaea'   Growth slow.

'Pyramidalis'   Growth conical.

'Redfield Seedling'   Growth slow, open; leaves short; reverts easily.

'Reinshaus'

'Sayville'   Growth compact, more rapid than 'Dove's Dwarf'.

'Sea Urchin'   Growth dwarf; leaves very short.

'Seacrest'   Small tree; growth conical, slender.

'Torulosa'   Branches twisted, the tips drooping; leaves twisted.

'Uconn'   Growth dwarf, dense, wide.

'Umbraculifera'   Growth slow, wide, umbrella-shaped.

'Uncatena'   Similar to 'Merrimack' but half the size.

'Variegata'   Leaves sometimes variegated yellow.

'Verkade's Broom'   Growth dwarf, rounded.

'White Mountain'   Leaves silver-blue.

'Winter Gold'   Same as 'Hillside Winter Gold'.

Pinus strobus **Cultivar Character Groupings**

Dwarf

'Bennett's Contorted'	'Green Shadow'	'Northway Broom'
'Blue Shag'	'Hillside Gem'	'Ontario'
'Brevifolia'	'Horsford'	'Uconn'
'Contorta Nana'	'Minima'	'Verkade's Broom'
'Densa'	'Minuta'	
'Dove's Dwarf'	'Nana'	

Rounded

'Anna Feile'	'Compacta'	'Horsford'	'Ontario'
'Bloomer's Dark Globe'	'Green Shadow'	'Minima'	'Umbraculifera'
'Bloomer's Globe'	'Greg's Form'	'Nana'	'Verkade's Broom'

27.   P. sylvestris **L.   Scotch pine.   Scots pine.**

Native to Europe and northern Asia. Cultivated as a Christmas tree, and many varieties planted as ornamentals. Grows rapidly in almost any location. Leaves 2 to a bundle, twisted, 4–8 cm (1½–3 in) long. Cones 4–6 cm (1½–2½ in) long.

'Albyn'   Growth dwarf, close to the ground.

'Argentea'   Leaves glaucous.

'Argentea Compacta'   Growth dwarf, conical or rounded; leaves glaucous.

'Aurea'   Growth slow; leaves yellow-green, becoming yellow.

'Aurea Nana'   Same as 'Aurea'.

'Aureopicta'   Growth dwarf; leaf tips mostly yellow.

'Bakony'   Growth conical, upright, although slow and squat in early years.

'Barrie Bergman'   Foliage with patches of white.

'Beauvronensis'   Growth dwarf, wider than high.

'Bennett Compact'   Same as 'Compact'.

'Bennett's Short-Leaf'

'Bergman'   Growth dwarf, dense, flat-topped.

'Columnaris Compacta'   Growth columnar; branches many, crowded.

'Compact'   Growth dwarf.

'Compressa'   Growth dwarf, columnar; leaves crowded, glaucous.

'Doone Valley'   Growth dwarf; leaves blue.

'Fastigiata'   Growth columnar; branches tending to be vertical.

'Genevensis'   Growth dwarf; branches few, far apart; leaf bundles in tufts.

'Glauca Compacta'   Growth dwarf; similar to 'Watereri'.

'Glauca Globosa'   Growth dwarf, rounded.

'Glauca Nana'   Growth dwarf, rounded; leaves blue.

'Globosa'   Growth dwarf, rounded, compact, dense; branches short.

'Globosa Virdis'   Growth dwarf, rounded; leaves green, very long.

'Gold Coin'   Growth dwarf; leaves yellow, stiff.

'Gold Medal'   Growth dwarf.

'Grand Rapids'   Growth dwarf.

'Greg's Variegated'   Foliage spotted with white or yellow.

'Guadraround'

'Hale's Prostrate'

'Heit's Pygmy'

'Helene Bergman'

'Hibernica'   Growth dwarf, rounded, open but dense.

'Hillside Creeper'   Growth spreading along ground.

'Hillside Weeper'   Branches drooping.

'Iceni'   Growth dwarf, with open, loose appearance.

'Inverleith'   Similar to 'Variegata'.

'Kamon Blue'   Growth conical; leaves blue.

'Kluis Pyramid'   Growth conical, dense.

'Lodgehill'   Growth dwarf.

'Mitsch Weeping'   Branches drooping.

'Mongolica'

'Monophylla'   Leaves of new growth appearing as one leaf in each bundle.

'Moseri'   Growth dwarf, rounded; foliage yellow in winter.

'Mount Vernon Blue'   Foliage blue.

'Nana'   Growth slow, compact; leaves short.

'Nana Compacta'   Growth dwarf, compact.

'Nisbet's Gem'   Growth dwarf, conical; leaves short, crowded.

'Pierson's Ridge'

'Pygmaea'   Growth dwarf, rounded, dense.

'Rependa'   Growth wide; branches thick, few; leaves spaced far apart.

'Repens'   Growth dwarf; buds resinous; short-lived.

'Riverside Gem'   Growth slow, rounded to conical, dense, sturdy.

'Rustic'   Branches somewhat twisted; leaves thick, twisted.

'Saxatilis'   Growth dwarf, compact; leaves blue.

'Sentinel'   Growth columnar, dense; branches ascending.

'Sherwood'   Growth dwarf, upright.

'Spaan's Slow Column'   Growth slow, columnar, dense; branches ascending.

'Spanish'   Growth wide, spreading.

'Tabuliformis'   Same as 'Nana Compacta'.

'Twiggy'   Growth rounded, open; branches numerous.

'Variegata'   Leaves with yellow or white patches.

'Virgata'   Branches few, long and snakelike in upper part of tree.

'Viridis Compacta'   Same as 'Globosa Viridis'.

'Watereri'   Growth dwarf, conical when young, becoming rounded; branches and leaves dense.

'Windsor'   Growth dwarf, bun-shaped; leaves small.

Pinus sylvestris Cultivar Character Groupings

Dwarf

'Albyn'	'Glauca Nana'	'Nana Compacta'
'Argentea Compacta'	'Globosa'	'Nisbet's Gem'
'Aureopicta'	'Globosa Viridis'	'Pygmaea'
'Beauvronensis'	'Gold Coin'	'Repens'
'Bennett Compact'	'Gold Medal'	'Saxatilis'
'Bergman'	'Grand Rapids'	'Sherwood'
'Compact'	'Greg's Variegated'	'Tabuliformis'
'Compressa'	'Hibernia'	'Viridis Compacta'
'Doone Valley'	'Iceni'	'Watereri'
'Glauca Compacta'	'Lodgehill'	'Windsor'
'Glauca Globosa'	'Moseri'	

Rounded

'Argentea Compacta'	'Glauca Globosa'	'Globosa'	'Hibernia'
	'Glauca Nana'	'Globosa Viridis'	'Moseri'

'Pygmaea'   'Viridis Compacta'   'Windsor'
'Riverside Gem'   'Watereri'

28. *P. tabulaeformis* Carr.   Chinese pine.

Native to China. Rarely planted tree or low, irregularly growing
shrub. Leaves in bundles of 2 or 3, more often 2, 5–15 cm (2–6 in)
long. Cones 4.0–6.5 cm (1½–2½ in) long.

29. *P. thunbergiana* Franco.   Japanese black pine.

Native to Japan and hardy in northeastern North America. Often
planted as an ornamental. Growth irregular, the main trunk and main
branches often forking. Leaves 2 to a bundle, stiff, 5–13 cm (2–5
in) long. Cones 5–8 cm (2–3 in) long. Seeds 7 mm (¼ in) long.

'Ara-kawa-sho'   Similar to 'Corticosa'.

'Aurea'   Some leaves yellow.

'Ban-sho-ho'   Growth dwarf, wide.

'Beni-jamone-kuromatsu'   Similar to 'Oculis-draconis' but leaves with
    pink rings in winter.

'Compacta'   Growth slow.

'Corticosa'   Bark corky; many named bonsai forms.

'Ganseki-matsu'   Similar to 'Corticosa'.

'Girardii Nana'   Growth dwarf, conical; leaves one-third as long as
    normal for the species.

'Globosa'   Growth rounded, dense.

'Iseli Golden'   Foliage variegated yellow.

'Janome-matsu'   Same as 'Oculis-draconis'.

'Kikko-sho'   Similar to 'Corticosa'.

'Ku-ja-ku'   Leaves spotted with pink.

'Kuro-bandai-sho'   Same as 'Globosa'.

'Nishiki'   Same as 'Corticosa'.

'Oculus-draconis'   Leaves with alternating green and yellow rings.

'Ogon-kuromatsu'   Same as 'Aurea'.

'Pendula'   Branches drooping.

'Shgarimatsu'   Same as 'Pendula'.

'Shidorematsu'   Same as 'Pendula'.

'Shirago-kuromatsu'   Same as 'Variegata'.
'Thunderhead'   Growth dwarf, wide; foliage dense.
'Tigrina'   Growth dwarf; leaves irregularly striped white.
'Tura-ku Kuromatsu'   Same as 'Tigrina'.
'Variegata'   Leaves all or patched white.
'Yatsubusa'   Growth dwarf.

+30.   *P. virginiana* Mill.   Virginia pine.   Scrub pine.   Jersey pine.

A small scraggly native of the southern United States extending
north to southern New York. Infrequently cultivated. Leaves 2 to a
bundle, twisted, 4–8 cm (1½–3 in) long. Cones 4–6 cm (1½–2½
in) long. Seeds 7 mm (¼ in) long.

'Nashawena'   Growth dwarf, spreading, open; branches congested.
'Pocono'   Growth slow, spreading; similar to 'Nashawena'.
'Watt's Golden'   Foliage yellow.

**31.   *P. wallichiana* A. B. Jacks.   Himalayan pine.   Weeping
pine.   Blue pine.**

Native to the Himalayas, marginally hardy in northeastern North
America. Tall, graceful, drooping pine. Leaves 5 to a bundle, soft
with bluish tinge, 15–20 cm (6–8 in) long. Cones 15–26 cm (6–10
in) long and stalked, the stalk 2.5–5.0 cm (1–2 in) long.

'Densa'   Growth conical, dense; leaves short.
'Frosty'   Growth rapid; foliage appearing frosted in fall and winter.
'Glauca'   Leaves blue.
'Nana'   Growth dwarf, bushy; leaves shorter.
'Silverstar'   Growth rounded to columnar; branches and leaves stiff.
'Umbraculifera'   Growth dwarf, rounded.
'Vernisson'   Hardier than species normal; branches on young plants
    upright.
'Zebrina'   Leaves with yellow horizontal band about 2.5 cm (1 in) from
    their tips.

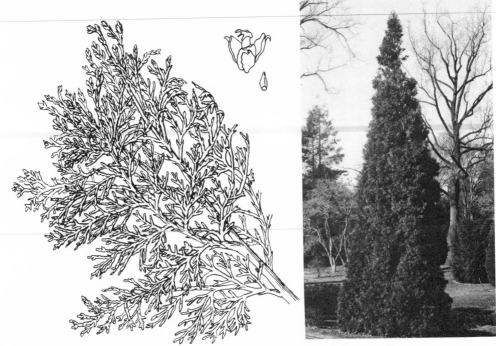

Platycladus orientalis

Platycladus Spach.   Oriental-arborvitae.

This genus contains only one species, which is described below.

1.   *P. orientalis* (L.) Franco.   Oriental-arborvitae.   Chinese-arborvitae.

Native to northern China and Korea. Until recently classified as *Thuja orientalis* L. or *Biota orientalis* (L.) Endl. Distinguished from *Thuja* by its lateral leaves, which are longer than the facial leaves, and its sprays, which tend to be in vertical rather than horizontal planes. Leaves without white markings, often with glands. Cones 1.3–2.5 cm (½–1 in) in diameter, with 4 or 6 scales. Seeds wingless, 2 per scale.

'Argenteovariegata'   Leaves at tips of young twigs white.

'Athrotaxoides'   Growth dwarf; branches thick; twigs thick; sprays not flat.

'Aurea'   Shrub; growth rounded, low, compact; leaves yellow,

140

becoming green.

'Aurea Compacta'   Same as 'Aurea Nana'.

'Aurea Conspicua'   Same as 'Conspicua'.

'Aurea Densa'   Same as 'Aurea Nana'.

'Aurea Globosa'   Same as 'Aurea Nana'.

'Aurea Nana'   Growth slow, rounded, then columnar; twigs more vertical; leaf tips yellow.

'Azurea'   Leaves glaucous.

'Bakeri'   Leaves pale green.

'Berckmans'   Similar to 'Aurea Nana'.

'Berckmans Golden'   Similar to 'Aurea Nana'

'Beverleyensis'   Growth columnar, compact; twig tips yellow.

'Blijdenstein'   Growth dwarf, conical, dense; branches crowded, erect.

'Blue Cone'   Growth rounded to conical, compact; leaves glaucous.

'Bonita'   Growth conical; leaf tips yellow.

'Brewers'

'Burtonii'

'Caesius'   Leaves glaucous.

'Chinensis'   Growth dwarf, rounded to conical.

'Columnaris'   Growth columnar.

'Compacta'   Same as 'Sieboldii'.

'Conspicua'   Growth slow, rounded, then columnar; branches more vertical; leaf tips yellow.

'Decussata'   Growth dwarf.

'Densa Glauca'   Growth rounded; foliage brown in winter.

'Dwarf Greenspike'   Growth dwarf, conical, compact; leaves needlelike.

'Elegantissima'   Growth conical; branches more vertical; leaf tips yellow, changing to red-brown.

'Excelsa'   Growth dwarf, conical, compact.

'Filifera'   Same as 'Flagelliformis'.

'Filiformis'   Same as 'Flagelliformis'.

'Filiformis Elegans'   Growth dwarf; leaves glaucous.

'Filiformis Erecta'   Growth dwarf; branches erect; twigs tightly clustered.

'Filiformis Nana'   Growth dwarf, congested.

'Flagelliformis'   Branches drooping and threadlike; leaves spaced far apart.

'Fruitlandii'   Growth dwarf, rounded; leaves dark green.

'Funiculata'   Twigs drooping; some leaves needlelike.

'Glauca'   Growth conical; leaves glaucous.

'Globosa'   Growth dwarf, rounded, compact.

'Golden Surprise'   Growth conical, wide, compact; leaves yellow, turning bronze.

'Goodwin'   Growth dense, symmetrical.

'Gracilis'   Growth slender, conical.

'Gracillima'   Growth rounded; leaves close together, dark green.

'Hillieri'   Similar to 'Aurea Nana'.

'Hohman'   Shrub; growth conical.

'Howardii'   Growth conical.

'Intermedia'   Twigs drooping; some leaves needlelike.

'Juniperoides'   Growth dwarf, columnar; branches crowded, numerous; leaves needlelike, glaucous.

'Macrocarpa'   Growth dwarf; twigs widely spaced.

'Magnifica'   Growth conical; branches evenly spaced, crowded.

'Maurieana'   Growth columnar.

'Mayhewiana'   Growth conical, compact; twigs tipped yellow.

'Meldensis'   Growth columnar.

'Millard's Gold'   Same as 'Aurea Nana'.

'Minima'   Shrub; growth dense and low.

'Minima Aurea'   Same as 'Aurea Nana'.

'Minima Glauca'   Growth dense, bushy.

'Monstrosa'   Growth dwarf; branches and foliage contorted.

'Nana'   Growth dwarf.

'Nana Compacta'   Growth dwarf, conical.

'Nepalensis'   Branches widely spaced, thin, spreading.

'Newarkii'

'Pumila Argentea'   Same as 'Summer Cream'.

'Pygmaea'   Growth dwarf.

'Pyramidalis'   Same as 'Stricta'.

'Pyramidalis Aurea'   Branches dense; leaves yellow, becoming yellow-green (not brown) in winter.

'Raffels'

'Ramsayi'

'Rosedale'   Same as 'Rosedalis'.

'Rosedalis'   Growth dwarf, rounded, compact; some leaves needlelike.

'Rosedalis Compacta'   Same as 'Rosedalis'.

'Sanderi'   Growth dwarf, rounded, low; leaves purple in winter.

'Semperaurescens' Growth dwarf, rounded; leaves yellow all year.

'Sieboldii' Growth dwarf, rounded, low; sprays crowded.

'Stricta' Growth conical.

'Summer Cream' Growth dwarf; young foliage white.

'Tatarica' Twigs yellow tipped.

'Tetragona' Growth dwarf, upright; branches four-sided.

'Texana Glauca' Growth conical; leaves glaucous.

'Threadleaf' Growth conical; branches threadlike, drooping.

'Triangularis' Growth dwarf.

'Westmont' Growth slow, rounded, spreading, weak; leaves with yellow tips.

Platycladus orientalis Cultivar Character Groupings

Dwarf

'Athrotaxoides'	'Fruitlandii'	'Pygmaea'
'Blijdenstein'	'Globosa'	'Rosedalis'
'Chinensis'	'Hillieri'	'Sanderi'
'Decussata'	'Juniperoides'	'Semperaurescen
'Dwarf Greenspike'	'Macrocarpa'	'Sieboldii'
'Excelsa'	'Monstrosa'	'Summer Cream'
'Filiformis Elegans'	'Nana'	'Tetragona'
'Filiformis Erecta'	'Nana Compacta'	'Triangularis'
'Filiformis Nana'	'Pumila Argentea'	

Columnar

'Aurea Nana'	'Blue Cone'	'Maurieana'
'Berckmans'	'Columnaris'	'Meldensis'
'Berckmans Golden'	'Conspicua'	
'Beverleyensis'	'Juniperoides'	

Rounded

'Aurea'	'Chinensis'	'Sanderi'
'Aurea Conspicua'	'Conspicua'	'Semperaurescens'
'Aurea Nana'	'Fruitlandii'	'Sieboldii'
'Berckmans'	'Globosa'	
'Berckmans Golden'	'Rosedalis'	

Yellow

'Aurea'	'Berckmans Golden'	'Mayhewiana'
'Aurea Conspicua'	'Beverleyensis'	'Pyramidalis Aurea'
'Aurea Nana'	'Conspicua'	'Semperaurescens'
'Berckmans'	'Hillieri'	'Tatarica'

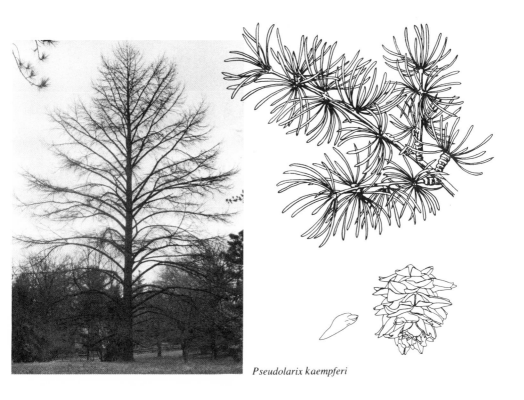

Pseudolarix kaempferi

Pseudolarix Gord.   Golden-larch.

This genus from China has a single species.

1. *P. kaempferi* Gord.   Golden-larch.

Native to eastern China. Valuable as an ornamental but slow growing and needing a well-drained soil and much space for its spreading
branches. Distinguished from the true larches by cones that hang
down from the branches and disintegrate at maturity, by its broader,
longer leaves, 4–5 cm (1½–2 in) long, and by its more prominently
ringed, longer knobs or spurs (short shoots) on the branches. Cones
4–6 cm (1½–3 in) long.

'Nana'   Growth dwarf.

145

Pseudotsuga menziesii

Pseudotsuga Carr.   Douglas-fir.

This genus from western North America and eastern Asia has about 10 species, only 1 of which is extensively cultivated.

1.   P. menziesii (Mirb.) Franco.   Douglas-fir.   Doug-fir.

One of the finest all-purpose trees. Unsurpassed among conifers in the quality of its construction lumber, a great quantity having been marketed from the vast forests in its native western North America. Makes a premium Christmas tree as well, although late spring frosts can be a problem. An excellent choice as a specimen tree, normally fast growing and shapely. Has soft but sharp-pointed leaves and requires little pruning. Hanging cones with protruding three-parted bracts produced at an early age. Leaves flat or angled, 2–3 cm (¾–1¼ in) long. Cones 5–10 cm (2–4 in) long.   var. *caesia* (Schwerin) Franco. Leaves glaucous; cones small.   var. *glauca* (Beissn.) Franco. Trees smaller; leaves shorter, thicker, bluish, glaucous; cones smaller with bracts reflexed. Seven cultivars of this botanical variety: 'Appressa', branches spreading; leaves pressed closely to twig, then spreading away from twig; 'Argentea', leaves silver; 'Argentea Pendula', twigs drooping; leaves glaucous; 'Compacta', growth conical, dense; leaves glaucous, short; 'Cripsa', leaves glaucous, wrinkled; 'Fletcheri', growth dwarf, low, spreading, flat-topped; leaves glaucous; 'Fretsii', leaves glaucous, short, wide.

'Albospica' Leaves on outer twigs nearly white when young.

'Anguina' Branches long; twigs few; leaves long.

'Astro Blue' Growth compact; foliage glaucous.

'Big Flats'

'Blue Wonder' Leaves on younger branches pointing forward, glaucous.

'Brevifolia' Tree smaller; leaves smaller.

'Caesia' Leaves glaucous.

'Densa' Growth dwarf, flat-topped, dense.

'Faberi' Leaves on outer branches yellow.

'Fastigiata' Growth columnar.

'Globosa' Growth dwarf, rounded.

'Graceful Grace' Branches drooping; foliage glaucous.

'Hess Select Blue' Growth conical; foliage glaucous.

'Hillside Gold' Foliage yellow.

'Hillside Pride' Growth dwarf, conical; branches more vertical.

'Holmstrup' Growth dwarf, conical, dense.

'Little Jon' Growth slow, conical, dense.

'Marshall' Growth dense.

'Mucronata Compacta' Growth dwarf, rounded.

'Nana' Growth slow, conical; branches numerous.

'Nidiformis' Growth dwarf, wider than high; middle branches flat, making a ''nest,'' or depressed area.

'Oudemansii' Growth conical; leaves rounded at the tips.

'Parkland Dwarf' Growth dwarf, low, flat.

'Pendula' Branches drooping; leaves dark green.

'Pendula Glauca' Branches drooping; foliage glaucous.

'Prostrata' Growth strong, low, spreading over the ground.

'Pumila' Growth dwarf, compact.

'Pyramidata' Growth slow, conical.

'Slavinii' Growth slow; leaves crowded, short.

'Tempelhof Compact'   Growth dwarf, dense, irregular, compact; buds numerous.

'Variegata'   Leaves on the upper side of branch yellow in summer.

'Viminalis'   Upper branches horizontal, with twigs hanging down.

'Viridis'   Leaves green; cones larger.

'Young's Broom'   Growth dwarf.

Pseudotsuga menziesii Cultivar Character Grouping

Dwarf

'Densa'	'Holmstrup'	'Pumila'
'Fletcheri'	'Mucronata	'Tempelhof
'Globosa'	Compacta'	Compact'
'Hillside Pride'	'Parkland Dwarf'	'Young's Broom'

Sciadopitys verticillata

Sciadopitys Sieb. & Zucc. Umbrella-pine.

This Japanese genus contains a single species.

1. *S. verticillata* Sieb. & Zucc. Umbrella-pine.

Native to Japan. A slow-growing, compact tree that needs a pro-tected, well-drained location to survive in the climate of north-eastern North America. The needlelike, flat leaves (actually each leaf consists of 2 leaves fused together, so that they appear to be 1) in whorls on the twig. Small scalelike leaves also present. Nee-dlelike leaves 5–13 cm (2–5 in) long. Cones upright, projecting horizontally, or hanging down, 4.5–10 cm (1¾–4 in) long.

'Helene Bergman'
'Knirps'   Growth dwarf, rounded, compact.
'Pendula'   Branches drooping.
'Pyramidalis Compacta'   Growth compact.
'Variegata'   One leaf of each pair yellow at branch ends.

149

Sequoia sempervirens

Sequoia Endl.   Redwood.

This genus is native to California and contains only one species.

1. *S. sempervirens* (D. Don) Endl.   Redwood.   Coast redwood.   Coastal redwood.   California redwood.

The famous coastal redwood of California and Oregon. Specimens sometimes more than 300 feet tall and including the tallest trees in the world. Few, if any, of its cultivars listed below hardy enough to persist in the climate of northeastern North America. Leaves needlelike (scalelike on young shoots), 0.7–2.5 cm (¼–1 in) long. Cones hanging down from the ends of the branches, 2–3 cm (¾–1⅛ in) long, 1.3–2.0 cm (½–¾ in) in diameter.

'Adpressa'   Growth dwarf, conical, crowded; leaves short, tips often white.

'Albospica'   Same as 'Adpressa'.

'Aptos Blue'   May be same as 'Soquel'.

'Argentea'

'Cantab' Same as 'Prostrata'.

'Compacta' Branches compact.

'Filiera Elegans' Branches horizontal; twigs cordlike; leaves becoming gradually smaller toward the end of the twigs.

'Glauca' Leaves glaucous.

'Gracilis' Twigs slender.

'Lawsoniana' Growth slower, compact, twigs short, stiff; leaves very short and stout.

'Los Altos'

'Majestic Beauty' Growth conical; leaves glaucous.

'Nana Pendula' Growth dwarf, spreading, open.

'Pendula' Branches drooping.

'Prostrata' Growth spreading over the ground; leaves wider.

'Repens' Same as 'Prostrata'.

'Santa Cruz' Growth conical.

'Soquel' Small tree; growth conical; foliage fine textured, lower surface retaining good color in winter.

'Taxifolia' Leaves wider.

'Variegata' Branches spreading; twigs short; leaves small, glaucous, often yellow or variegated yellow.

Sequoiadendron gigantea

Sequoiadendron Buch.　Giant-sequoia.

This California genus includes only one species.

1.　*S. gigantea* (Lindl.) Buch.　Giant-sequoia.　Giant-redwood.　Bigtree.　Sequoia.　Sierra redwood.

Ancient massive trees of California that unfortunately are only marginally hardy in northeastern North America. Grows well for a number of years in protected areas but is destroyed by hard winters. Leaves scalelike, 3–13 mm (⅛–½ in) long. Cones 4–8 cm (1¾–3 in) long.

'Argentea'　Leaves on young shoots with silver patches.

'Aurea'　Leaves on young shoots yellow.

'Compacta'　Growth thin; branches short; leaves pressed tightly to the twig.

'Glauca'　Leaves on young shoots glaucous.

'Pendula'　Branches drooping.

'Pygmaea'　Growth dwarf, bushy.

Taxodium distichum

Taxodium L. Rich. Baldcypress. Yew-cypress.

This genus has three species that are native to southern North America, one of which is often planted for ornament and one of which is very rarely cultivated (*T. mucronatum*). The leaves and short lateral branches are deciduous, which distinguishes this genus and *Metasequoia* from other genera. The leaves are solitary, alternate (in contrast to opposite leaves in *Metasequoia*), and short. The bark is scaly and shredding. Female cones are small, with winged seeds under each scale.

132. *T. distichum*

1. Leaves spreading away from the twig; twigs horizontal
   (Fig. 132) . . . . *T. distichum*
   Baldcypress
1. Leaves pressed next to and parallel to the twig, incurved; twigs
   upright (Fig. 133) . . . . *T. ascendens*
   Pond-cypress

133. *T. ascendens*

153

1. *T. ascendens* **Brongn.   Pond-cypress.**

Native to the southern states but hardy in northeastern North America. Leaves narrow, sharp-pointed, pressed close to and lying parallel to the upright twigs. Leaves 1.3–2.0 cm (½–¾ in) long. Cones 2.5 cm (1 in) in diameter.

2. *T. distichum* **(L.) L. Rich.   Baldcypress.**

Native as far north as Delaware but hardy in cultivation in northeastern North America. Leaves spreading away from the twig. Branches slightly drooping or hanging, forming a feathery, handsome tree. Leaves 2.5 cm (1 in) long. Cones 13 mm (½ in) in diameter.

'Nutans'
'Pendens'   Twigs drooping.
'Pendula'

Taxus cuspidata

Taxus ×media

Taxus L.   Yew.

The 10 species of yew, all native to the Northern Hemisphere, are some of the most useful conifers for landscape plantings. They are used frequently for hedges and for low plantings next to buildings. Caution should be exercised when they are incorporated into landscape design, because the dark green foliage may give a somber effect in an overabundance of yew plantings. The branches, leaves, and seeds when eaten are poisonous to humans and to domesticated livestock. Deer, however, browse them heavily. Most yews are shrubs, although some can attain enough height to be considered trees. The yews are characterized by dark green leaves, usually yellow-green on the lower surface, and by a peculiar red or sometimes yellow pulpy female reproductive structure that superficially resembles a berry.

This abundantly cultivated genus is unfortunately one of the most difficult to understand taxonomically. For horticulturalist and taxonomist alike it is extremely difficult to separate the species and hundreds of cultivars from each other. *Taxus* is distinguished from

155

134. *T. baccata*

135. *T. baccata*

136. *T. ×media*

137. *T. celebica*

138. *T. ×media*

139. *T. canadensis*

Cephalotaxus by its shorter, alternate leaves and by its red rather than purple female reproductive structures. *Taxus* differs from *Torreya* in its obviously alternate leaves and red female reproductive structures. The berrylike female reproductive structures distinguish *Taxus* from other conifers, which usually bear woody cones.

1. Bud scales blunt or with dull tip, only slightly keeled (Fig. 134).
   2. Leaves with long, tapering tips (Fig. 135) . . . . *T. baccata**
       English yew
   2. Leaves abruptly short-pointed (Fig. 136).
      3. Bud scales mostly not remaining attached (Fig. 137); leaves with midrib only slightly raised, short stalked, spreading . . . . *T. celebica*
       Chinese yew
      3. Bud scales remaining attached (Fig. 138); leaves with midrib raised . . . . *T. ×media**
       Hybrid yew
1. Bud scales often sharp-pointed, keeled (Fig. 139).
   4. Leaves 1.5–2.0 mm wide, flat, with slightly raised midrib.
      5. Leaf stalks green, shorter than 1 mm, barely evident; leaf edges slightly rolled under . . . . *T. canadensis+*
       Canadian yew
      5. Leaf stalks yellow, slender, longer than 1 mm; leaf edges flat . . . . *T. brevifolia*
       Pacific yew
   4. Leaves 2–3 mm wide, with prominent midrib. (It is impossible, in most cases, to differentiate adequately between *T. cuspidata* and its two hybrids, *T. ×media* and *T. ×hunnewelliana* without seeing the plants side by side.)
      6. Twigs shining; bud scales keeled; leaves with wide, yellowish longitudinal bands on lower surface.
         7. Midrib prominent, well raised . . . . *T. cuspidata**
           Japanese yew
         7. Midrib not as elevated . . . . *T. ×media**
           Hybrid yew
      6. Twigs less shining, more slender; bud scales narrower, less keeled; leaves narrower, with more abruptly tapered tips, not as yellow on lower surface . . . . *T. ×hunnewelliana*
       Hybrid yew

*1. *T. baccata* L. English yew. European yew.

Native to Europe. Can be a tree to 22 m (70 ft). Densely branched. Leaves 3 cm (1¼ in) long, with prominent midrib. Some of the many European cultivars not cultivated in the United States included in the list below.

'Adpressa'   Leaves with very small, abrupt tips.

'Adpressa Aurea'   Leaves with abrupt tips, yellow at branch tips.

'Adpressa Erecta'   Growth conical; main branches erect.

'Adpressa Fowle'   Growth compact, wide.

'Adpressa Pyramidalis'   Growth conical; main branches erect.

'Adpressa Variegata'   Growth slow; leaves pale yellow.

'Amersfoort'   Growth dwarf, stiff; twigs few; leaves numerous, slightly convex.

'Argentea'

'Argentea Minor'   Same as 'Dwarf White'.

'Aurea'   Growth compact; leaves yellow, changing to green.

'Aureovariegata'   Leaves with some yellow coloring.

'Aurescens'   Foliage yellow.

'Brevifolia'   Leaves many, crowded, small.

'Buxifolia'

'Cavendishii'   Growth low, spreading; twig tips drooping.

'Cheshuntensis'   Growth wide, spreading; leaves glaucous.

'Columnaris'   Growth columnar, dense.

'Compacta'   Growth dwarf, conical, compact; leaves curved, short.

'Corona'   Growth dwarf, flat-topped.

'Davisii'   Growth conical; branches and twigs thin, crowded.

'Decora'   Growth dwarf, dense, low, and spreading.

'Dovastonii'   Branches spreading; twigs drooping.

'Dovastonii Aurea'   Twigs drooping; leaves with patches of yellow.

'Dovastonii Aureovariegata'   Twigs drooping; leaves with patches of yellow.

'Dwarf White'   Growth slow, low; leaves with white markings.

'Elegantissima'   Growth compact; leaves striped yellow.

'Epacrioides'   Growth dwarf, upright; branching irregular.

'Erecta'   Growth upright, bushy.

'Erecta Aurea'   Growth upright, bushy; foliage yellow.

'Ericoides'   Growth dwarf, low, spreading; leaves small, bronze in winter.

'Expansa'   Growth spreading.

'Fastigiata'   Growth columnar; leaves darker green.

'Fastigiata Aurea'   Growth columnar; leaves yellow.

'Fastigiata Aureomarginata'   Growth columnar; foliage with some yellow.

'Fastigiata Aureovariegata'   Growth columnar; foliage patched yellow.

'Fastigiata Nana'   Growth dwarf, columnar.

'Fowle'   Same as 'Adpressa Fowle'.

'Fructo-luteo'   Reproductive structure yellow.

'Glauca'   Leaves glaucous on lower surface.

'Green Silver'   Same as 'Silver Green'.

'Gracilis Pendula'   Branches slender, drooping.

'Hessei'   Growth upright; branches many, crowded; leaves long, wide.

'Hibernica'   Growth columnar.

'Horizontalis'   Growth low, lacking leading shoot; branches wide spreading.

'Imperialis'   Growth dense, upright.

'Jacksonii'   Branches spreading, drooping at tips.

'Kadett'   Growth conical; branches upright.

'Knirps'   Growth dwarf; twigs dark brown.

'Lutea'   Reproductive structure yellow.

'Melfard'   Growth columnar; leaves small.

'Michelii'   Growth low, dense.

'Nana'   Growth dwarf, dense.

'Neidpathensis'   Growth vigorous; much-branched.

'Nigra'   Twigs slender turning reddish in winter sun, tips drooping.

'Nutans'   Growth dwarf; branches ascending; foliage dense.

'Overeynderi'   Growth upright.

'Page'   Growth dwarf; foliage dense.

'Paulina'   Growth dwarf, conical, dense.

'Pendula'   Branches slender, drooping.

'President'   Growth dwarf.

'Procumbens'   Growth dwarf; branches many, lying close to the ground.

'Prostrata'   Growth dwarf, spreading.

'Pseudo-procumbens'   Growth low; branches elongated, numerous, drooping.

'Pumila'   Growth dwarf.

'Pumial Aurea'   Growth dwarf, rounded; young shoots yellow; leaves yellow.

'Pygmaea'   Growth dwarf, rounded to conical.

'Pyramidalis'   Growth upright, bushy.

'Raket'   Growth conical, dense.

'Recurvata'   Branches horizontal; leaves recurved.

'Repandens'   Growth dwarf; branches spreading close to ground; leaves glaucous.

'Repandens Aurea'   Growth dwarf; branches close to ground; foliage yellow.

'Repens'   Growth dwarf, spreading.

'Repens Aurea'   Same as 'Repandens Aurea'.

'Semperaurea'   Leaves yellow.

'Silver Green'   Growth compact.

'Standishii'   Growth dwarf, columnar; leaves yellow.

'Stricta'   Growth columnar; leaves darker green.

'Summergold'   Leaves yellow.

'Tardiva'   Leaves with very small abrupt tips.

'Variegata'   Leaves with patches of white.

'Washingtonii'   Growth wide, spreading; leaves yellow.

Taxus baccata Cultivar Character Groupings

Dwarf

'Amersfoort'	'Ericoides'	'Paulina'	'Pygmaea'
'Compacta'	'Fastigiata Nana'	'President'	'Repandens'
'Corona'	'Knirps'	'Procumbens'	'Repandens Aurea'
'Decora'	'Nana'	'Prostrata'	'Repens'
'Elegantissima'	'Nutans'	'Pumila'	'Standishii'
'Epacrioides'	'Page'	'Pumila Aurea'	

Low, Spreading over the Ground

'Cavendishii'	'Elegantissima'	'Procumbens'	'Repandens'
'Decora'	'Ericoides'	'Prostrata'	'Repandens Aurea'

Columnar

'Columnaris'	'Fastigiata Aureomarginata'	'Hibernica'
'Elegantissima'		'Melfard'
'Fastigiata'	'Fastigiata Aureovariegata'	'Standishii'
'Fastigiata Aurea'	'Fastigiata Nana'	'Stricta'

Yellow

'Adpressa Aurea'	'Aurea'	'Aurescens'
'Adpressa Variegata'	'Aureovariegata'	'Dovastonii Aurea'

'Dovastonii Aureovariegata'	'Fastigata Aureovariegata'	'Standishii'
'Elegantissima'	'Pumila Aurea'	'Summergold'
'Erecta Aurea'	'Repandens Aurea'	'Washingtonii'
'Fastigiata Aurea'	'Repens Aurea'	
'Fastigiata Aureomarginata'	'Semperaurea'	

2. *T. brevifolia* Nutt.   Pacific yew.   Western yew.

A native tree or shrub of western North America. Not very common or successful in cultivation in northeastern North America. Leaves thin, 7–17 mm (¼–¾ in) long, with raised midrib and yellowish leaf stalks.

'Erecta'   Growth columnar.
'Nana'   Growth dwarf.
'Nutallii'   Branches drooping.
'Pyramidalis'

+3. *T. canadensis* Marsh.   Canadian yew.   Ground hemlock.   American yew.

The only species of yew native to northeastern North America. A shrub that is low growing and straggly in the wild. Very hardy but relatively undesirable aesthetically. Leaves dense on the branches, 1–2 cm (⅜–¾ in) long, with raised midrib on both surfaces.

'Aurea'   Leaves with some yellow.
'Compacta'   Branches more dense.
'Dwarf Hedge'   Growth dwarf; branches stiff, upright.
'Fastigiata'   Growth dwarf, erect.
'Pyramidalis'   Same as 'Stricta'.
'Stricta'   Growth dwarf, erect.
'Washingtonii'

4. *T. celebica* (Warburg) Li.   Chinese yew.

A shrub that can be a small tree in its native China. Leaves 1.2–3.0 cm (½–1⅛ in) long, spaced farther apart on the branches than in most other yews. Leaves short stalked and with only a slightly raised midrib. Not common. Known mostly as *T. chinensis* Rehd.

*5. *T. cuspidata* **Sieb. & Zucc.** **Japanese yew.**

Can be a small tree, but this native of Japan is known in North America mostly as a small shrub or bush with a single trunk. Leaves short stalked, 1.2–2.5 cm (½–1 in) long, with prominent midrib. Planted abundantly in northeastern North America.

'Adams'   Growth columnar; leaves forming pronounced "V" on twigs.
'Andersonii'   Leaves larger.
'Aristocrat'   Twigs abundant, slender, with nodding tips.
'Aurea'   Leaves with some yellow.
'Aurescens'   Growth low; young leaves yellow.
'Barnes'
'Bobbink'   Growth dwarf, conical, dense.
'Bright Gold'   Growth low, spreading; foliage yellow.
'Brownii'   Growth compact, spreading.
'Buffumii'
'Bulkii'   Growth erect; leaves dark green.
'Capitata'   Growth spreading or conical and erect.
'Cherry Hill'
'Columnaris'   Growth columnar.
'Columnaris Compacta'
'Compacta'   Growth slow, dense.
'Corliss Special'
'Densa'   Growth low, dense.
'Densiformis'   Growth spreading, dense, compact.
'Depressa'   Growth low; branches close to the ground.
'Dwarf Bright Gold'   Same as 'Bright Gold'.
'Erecta'   Growth columnar.
'Expansa'   Growth spreading.
'Fastigiata'   Growth dwarf, columnar; leaves yellow.
'Fieldsii'   Growth spreading, low, compact; leaves small.
'Giraldii'
'Green Wave'   Growth low; branches arching.
'Heasleyi'
'Hillii'   Growth conical; leaves dark green.

'Hitii'   Growth columnar or conical, compact; branches erect.

'Hoytii'   Growth upright but spreading, compact; leaves close together.

'Intermedia'   Growth slow, rounded, dense.

'Luteobaccata'   Reproductive structure yellow.

'Midget'   Same as 'Bobbink'.

'Minima'   Growth dwarf; leaves dark green.

'Nana'   Growth slow, dense.

'Nana Compacta'

'Nana Grandifolia'

'Nana Pyramidalis'   Growth conical; leaves crowded.

'Nigra'   Leaves very dark green.

'Ovata'   Leaves wide.

'Prostrata'   Growth wide, spreading, close to the ground.

'Pyramidalis'   Growth columnar.

'Robusta'   Growth vigorous, with either erect or spreading branches.

'Rustique'   Branches long; some leaves long, wide.

'Sieboldii'   Growth dense; foliage glaucous.

'Stovekenii'   Growth columnar, vigorous.

'Stricta'   Growth low; branches irregular.

'Taunton'   Same as *T.* ×*media* 'Taunton'.

'Thayeri'   Growth wide, spreading; branches horizontal, feathery.

'Vermeulen'   Growth slow, upright, vase-shaped.

'Viridis'

'Visseri'   Growth slow, spreading, dense.

'Wilsonii'   Growth vase-shaped, compact; leaves dark green.

6.   *T.* ×*hunnewelliana* Rehd.   (*T. cuspidata* Sieb. & Zucc. × *T. canadensis* Marsh.)   Hybrid yew.   Hunnewell yew.

A cross between Japanese and Canadian yews. Not commonly cultivated in the Northeast. Similar to Japanese yew but with narrower leaves.

'Richard Horsey'   Growth dwarf, wide; branches many, crowded.

*7.   *T.* ×*media* Rehd.   (*T. cuspidata* Sieb. & Zucc. × *T. baccata* L.) Hybrid yew.   Anglo-Japanese yew.

A cross between Japanese and English yew frequently seen in nurseries and in plantings in northeastern North America. Leaves wider and stouter, with the midrib more raised than in English yew. Young twigs remaining green for two years instead of one, in contrast to Japanese yew.

'Amherst'   Growth spreading.

'Andersonii'   Growth erect but freely branching.

'Andorra'   Growth conical; branches erect.

'Anthony Wayne'   Growth columnar, vigorous; foliage yellow.

'Beanpole'

'Berryhill'   Growth low, wide, compact, dense.

'Brevicata'   Growth compact; leaves short, wide.

'Brevimedia'   Plant hardy; growth spreading, then upright; leaves dark green.

'Broad Beauty'   Growth low, spreading.

'Brownhelm'   Growth vigorous; twigs numerous.

'Brownii'   Growth conical; leaves short, close together.

'Burr'   Growth dense; leaves dark green.

'Chadwick'   Growth low; branches with some unsweep.

'Cliftonii'   Growth erect, dense; leaves dark green.

'Cole'   Growth low, spreading, dense.

'Coleana'   Growth wide spreading; some branches erect.

'Columnaris'   Growth columnar; leaves dark green.

'Compacta'   Growth slow, spreading.

'Costich'   Growth columnar.

'Dark Green Spreader'   Growth spreading, compact.

'Densiformis'   Growth spreading, dense; leaves dark green.

'Devermanni'   Growth dense, compact; leaves waxy.

'Done Well'   Growth compact, with flaring terminal branches.

'Drulia'   Reproductive structure split into two or three parts.

'Dutweileri'   Growth vase-shaped; leaves crowded.

'Emerald'   Growth vigorous.

'Erecta'   Growth tall, somewhat columnar.

'Everflow'   Growth low, spreading.

'Fairview'   Growth dwarf.

'Fastigiata'

'Flemer'   Growth slow, rounded, compact.

'Flushing'   Growth dwarf, columnar; branches stout, erect.

'Gem'   Growth rapid, rounded, dense.

'Grandifolia'   Growth compact, upright; leaves large, dark green.

'Green Candle'   Growth columnar; branches stiff.

'Green Mountain'   Growth compact; twigs twisted.

'Halloriana'   Growth conical.

'Hatfieldii'   Growth conical; leaves widely spaced.

'Heasleyi'   Growth dwarf.

'Helleri'   Growth rapid, erect.

'Henryi'   Growth spreading, dense; branches becoming erect.

'Hetz'

'Hicksii'   Growth columnar; branches upright; very commonly planted.

'Hill'   Growth conical to rounded.

'H. M. Eddie'   Growth conical; foliage dark green.

'Hoogendorn'   Growth spreading to upright.

'Hummeri'

'Kallay'

'Kelseyi'   Growth upright, dense.

'Kobel'   Growth wide, spreading, loosely branched.

'Kohli'   Growth rounded, upright; leaves very dark green.

'Lodi'   Growth compact but leaves loosely arranged on terminal shoots.

'Matinecock'

'Microphylla'   Leaves small.

'Mitiska Upright'   Growth rectangular, upright.

'Moon'   Growth dense; branches becoming erect.

'Nana-grand'   Growth rounded, compact

'Natorp'   Growth dense, compact.

'Newport'   Growth slow; foliage very dense, forming "mounds."

'Nidiformis'   Growth wide, spreading, low, with depressed area in
    center of the plant.

'Nidiformis Nigra'   Same as 'Nidiformis'.

'Nigra'   Growth compact; leaves very dark green.

'Ohio Globe'   Growth rounded.

'Ovata'   Growth wide, upright; leaves large, dark green.

'Parade'   Growth columnar but wide; leaf tips dark green.

'Peterson'   Growth wide, spreading.

'Pilaris'   Growth slow, columnar, narrow.

'Pyramidalis'   Growth slow, columnar; branches widely spaced.

'Robusta'   Growth columnar, compact.

'Roseco'   Growth wide, spreading; foliage abundant, dark green.

'Runyan'   Growth compact.

'Sebian'   Growth low, becoming wide, spreading, compact.

'Sentinalis'   Growth slow, columnar; branches and leaves widely
    spaced.

'Stovekenii'   Growth columnar; vigorous; branches erect.

'Straight Hedge'   Growth strong, upright, branches numerous.

'Stricta'   Growth erect, compact.

'Taunton'   Growth rounded, dense.

'Thayeri'   Growth slow, vase-shaped, with few side branches.

'Totem'   Growth columnar, becoming wider with age; twigs pressed to the branch, short.

'Vermeulen'   Growth slow, columnar, dense; branches erect.

'Verticalis'

'Viridis'   Growth columnar, narrow, dense; leaves twisted.

'Wardii'   Growth erect, compact.

'Wellesleyana'   Growth columnar or vase-shaped; leaves thick.

'Wilsonii'   Growth rounded, compact.

'Wiltonii'   Growth bushy, wide.

'Wymanii'   Growth vase-shaped; branches close together.

Taxus × *media* **Cultivar Character Grouping**

Columnar

'Anthony Wayne'	'Flushing'	'Pilaris'	'Totem'
	'Green Candle'	'Pyramidalis'	'Vermeulen'
'Columnaris'	'Hicksii'	'Robusta'	'Viridis'
'Costich'	'Parade'	'Sentinalis'	'Wellesleyana'
'Erecta'			

Thuja occidentalis

Thuja L.   Arborvitae.

140. *P. orientalis*

141. *T. plicata*

This eastern Asian and Northern American genus contains five species, one native to northeastern North America. A sixth, *T. orientalis*, is now placed in the genus *Platycladus*. Arborvitae is frequently cultivated alongside buildings, in specimen plantings, and in hedges and windbreaks. Leaves are small, scalelike (needlelike on young foliage), entire, overlapping, and opposite. The genus is distinguished from *Platycladus* by its lateral leaves, which are only as long as they are wide, and from *Chamaecyparis* and *Cupressus* by its overlapping cone scales and flattened sprays. Species of *Chamaecyparis* with very flat sprays are difficult to distinguish from *Thuja* vegetatively; the two genera are usually distinguishable by the sharper-pointed leaf tips of *Chamaecyparis*.

1. Sprays in various planes, tending to be parallel to trunk or main stem; lateral leaves longer than wide (Fig. 140) . . . .*T. orientalis*
(see *Platycladus orientalis*)
Oriental-arborvitae

1. Sprays mostly in horizontal planes; lateral leaves as long as wide (Fig. 141).

166

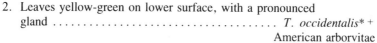

142. *T. plicata*

2. Leaves yellow-green on lower surface, with a pronounced
   gland .... *T. occidentalis** +
                                                            American arborvitae
2. Leaves with triangular white markings on lower surface, with or
   without a gland.
   3. Leaves on main axes of branches ending in long tapering points
      (Figs. 142 and 143) .... *T. plicata*
                                                            Western red-cedar
   3. Leaves on main axes of branches ending in short, triangular,
      somewhat spreading points (Figs. 144 and 145) ..... *T. standishii*
                                                            Japanese arborvitae

143. *T. plicata*

* + 1. *T. occidentalis* L.   American arborvitae.

144. *T. standishii*

One of the most frequently planted conifers in northeastern North America. A native of eastern North America that forms dense stands in wet areas. Excellent for planting in hedges or close to buildings but can also make a good specimen standing alone. Leaves bearing glands and lacking the white markings on the lower surface that are evident in some other species. Cones 13 mm (½ in) long, with 8–10 scales. Seeds winged.

145. *T. standishii*

'Ada'   Growth columnar.

'Alba'   Growth conical; twigs and sprays white tipped.

'Americana'

'Aurea'   Growth rounded; leaves yellow.

'Aurea Americana'   Leaves on young shoots yellow.

'Aureovariegata'   Foliage variegated yellow.

'Aurescens'   Foliage on young twigs yellow.

'Batemannii'   Growth dwarf.

'Beaufort'   Young leaves white.

'Beteramsii'   Leaves on outer twigs reddish in spring.

'Bluespire'

'Bodmeri'   Growth dwarf; branches short.

'Boothii'   Growth dwarf, rounded, becoming flat-topped, compact; leaves large.

'Brandon'   Growth columnar, dense, compact.

'Brewer's Hybrid'

'Brubakeri'

'Buchananii'   Branches thin, widely spaced; twigs widely spaced, short.

'Burrowii'   Leaves yellow.

'Caespitosa'   Growth dwarf, bun-shaped, low; foliage congested, irregular.

'Canadian Green'   Growth rounded.

'Caucasia'

'Cheshire'

'Cloth of Gold'   Shrub; growth slow; leaves yellow.

'Columbia'   Growth columnar; twig tips white.

'Columna'   Growth narrowly columnar.

'Compacta'   Growth dwarf; branches crowded.

'Compacta Americana'   Growth conical, compact.

'Compacta Erecta'   Growth conical; branches erect; leaves dark green.

'Conica'   Growth conical.

'Conica Densa'   Growth rounded, dense.

'Cristata'   Growth dwarf, rounded, upright; sprays curved.

'Cristata Aurea'   Growth dwarf, rounded; sprays curved; foliage yellow-green.

'Danica'   Growth dwarf, rounded, compact; similar to 'Woodwardii'.

'Dark American'

'Degroot's Emerald Spire'

'Densiforma'   Growth very dense.

'Dirigo Dwarf'   Growth dwarf.

'Dorsett Yellow'

'Douglasii Aurea'   Growth conical, slender; leaves yellow, bronzed in winter.

'Douglasii Pyramidalis'   Growth columnar, dense or slender and open; foliage twisted.

'Dumosa'   Growth dwarf, dense; branches short, fan-shaped, numerous.

'Elegantissima'   Growth conical; leaf tips white, becoming brown.

'Ellwangeriana'   Leaves needlelike on many of the branches.

'Ellwangeriana Aurea'   Growth dwarf, rounded, leaves yellow, some needlelike.

'Emerald'   Growth conical; foliage dense, dark green; same as 'Smaragd'.

'Emerald Green'   Same as 'Smaragd'.

'Endean'   Growth conical, open.

'Erecta'

'Ericoides'   Growth conical, low, dense.

'Europe Gold'   Foliage yellow; similar to 'Fastigiata'.

'Fastigiata'   Growth conical; twigs crowded.

'Filicoides'   Growth dense; twigs short.

'Filiformis'   Growth dwarf, rounded; twig tips threadlike, curving.

'Frieslandia'

'Froebellii'   Growth dwarf.

'George Peabody'   Foliage yellow.

'George Washington'   Growth conical; leaves with yellow patches.

'Giganteoides'   Growth rapid, vigorous; branches stiff, thick; twigs widely spaced.

'Globosa'   Growth dwarf, rounded.

'Globosa Aurea'   Growth dwarf, rounded; foliage yellow; probably same as 'Golden Globe'.

'Globosa Nana'

'Globosa Rheindiana'

'Globularis'

'Goldcrest'   Growth conical; tips of young leaves yellow.

'Gold Dust'

'Golden'   Growth slow, upright; foliage yellow.

'Golden Champion'

'Golden Globe'   Growth dwarf, rounded; leaves yellow.

'Green Harbor'

'Green Midget'   Growth dwarf, rounded.

'Hetz Junior'   Growth conical; leaves may be needlelike; similar to 'Ericoides'.

'Hetz Midget'   Growth dwarf, rounded.

'Hetz Midget Variegated'   Growth dwarf; leaves with patches of yellow.

'Hetz Wintergreen'   Growth vigorous; leaves remaining green.

'Hollandica'   Growth dense.

'Holmstrup'   Growth conical, compact; leaves remaining green.

'Holmstrup Yellow'   Leaves yellow.

'Hookeriana'

'Hoopesii'   Growth conical; leaves crowded.

'Hoseri'   Growth dwarf, rounded.

'Hoveyi'   Growth dwarf; rounded; leaves brown in winter.

'Hudsonica'   Growth rounded, dense.

'Indomitable'   Hardy; leaves with some reddish brown color in winter.

'Intermedia'   Growth dwarf, compact.

'Leptoclada'   Growth dwarf.

'Little Champion'   Growth dwarf though rapid at first; sprays lacy.

'Little Gem'   Growth dwarf, rounded, becoming conical, dense; twigs crimped.

'Little Giant'   Growth slow, rounded, compact.

'Lombarts Dwarf'   Growth dwarf, rounded, dense.

'Lombarts Wintergreen'   Leaves green all year.

'Lutea'   Growth conical; sprays yellow.

'Lutea Nana'   Growth dwarf, conical; sprays yellow; similar to 'Ellwangeriana Aurea'.

'Lutescens'   Twigs stout; foliage yellow.

'Lycopodiodes'

'Malonyana'   Growth columnar, dense, compact.

'Martinius'

'Mastersii'   Branches short, stiff; leaves crowded.

'Meckii'   Growth rounded.

'Meinekes Zwerg'   Growth dwarf, rounded, dense; spray tips white.

'Milleri'   Sprays lacy, narrow, in random directions but held upright.

'Minima'   Same as 'Pygmaea'.

'Nana'   Growth dwarf, compact.

'Nigra'   Leaves dark green.

'Ohlendorfii'   Growth dwarf, irregular; young twigs on lower branches with needlelike leaves.

'Pendula'   Branches drooping.

'Pulcherrima'

'Pumila Sudworthii'   Growth dwarf, rounded; leaves yellow at branch tips.

'Pygmaea'   Growth dwarf, stunted; branches irregular.

'Pyramidalis'   Growth conical, hardy.

'Pyramidalis Compacta'   Growth narrowly conical, dense.

'Pyramidalis Douglasii'

'Queen Victoria'

'Recurva'

'Recurva Nana'   Twigs turning back at end of branches.

'Recurvata'

'Reedii'   Growth dwarf, rounded.

'Reidii'   Growth dwarf, wide.

'Rheindiana'   Growth dwarf, rounded.

'Rheindiana Globosa'

'Rheingold'   Growth dwarf, rounded, becoming conical; leaves yellow; similar to 'Ellwangeriana Aurea'.

'Rheingold Beattie' Growth conical; foliage yellow, turning bronze in winter.

'Riversii' Growth conical; leaves yellow.

'Robusta' Same as 'Wareana'.

'Robusta Recurva'

'Rosenthalii' Growth conical, compact, slow; sprays in more vertical planes.

'Semperaurea' Growth conical, dense; leaves at least tipped with yellow.

'Sherman' Growth conical; leaves dark green.

'Sherwood Column' Growth columnar.

'Sherwood Frost' Growth upright; foliage spotted white.

'Sherwood Moss' Growth dwarf, conical.

'Skogholm' Growth columnar, dense.

'Smaragd' Growth conical, compact.

'Smithiana' Growth low, compact; foliage purple in fall.

'Sphaerica' Growth dwarf, rounded, becoming upright; sprays curled, crimped; leaves tiny.

'Spiralis' Growth conical; branches short; twigs spiral, curved.

'Sudworthii' Foliage yellow.

'Sunkist' Growth dwarf, conical; leaves yellow.

'Tatarica'

'Techny' Growth slow, conical; leaves green all year.

'Theodonensis' Twigs thick; foliage dark green.

'Thujopsoides' Branches widely spaced; twigs thick, bending down.

'Tiny Tim' Growth dwarf, rounded.

'Tom Thumb' Similar to 'Ellwangeriana'.

'Umbraculifera' Growth dwarf, rounded, compact; leaves somewhat glaucous.

'Unicorn' Growth columnar; foliage dark green.

'Van der Bom' Growth dwarf, rounded when young.

'Variegata' Foliage variegated white.

'Vervaenaena' Growth conical; twigs crowded; leaves varying from green to yellow.

'Virescens'

'Wagneri' Growth rounded, compact.

'Wansdyke Silver' Growth conical, dense; leaves with patches of white.

'Wareana' Growth slow, conical; twigs often vertical.

'Wareana Globosa'   Growth dwarf, rounded, congested.

'Wareana Lutescens'   Similar to 'Wareana'; leaves yellow.

'Washingtonia Aurea'

'Watereri'   Growth dwarf; twigs thin and fine, often spiraled; leaves sometimes needlelike.

'Watnong Gold'   Growth columnar, compact; leaves yellow.

'Waxen'   Sprays nodding; leaves yellow-green.

'Westminster'

'Wintergreen'   Growth columnar or conical.

'Wintergreen Pyramidal'

'Woodwardii'   Growth dwarf, rounded; leaves dark green, turning brown in winter.

Thuja occidentalis Cultivar Character Groupings

Dwarf

'Batemannii'	'Globosa Aurea'	'Pumila Sudworthii'
'Bodmeri'	'Golden Globe'	'Pygmaea'
'Boothii'	'Green Midget'	'Reedii'
'Caespitosa'	'Hetz Midget'	'Rheindiana'
'Compacta'	'Hetz Midget Variegated'	'Rheingold'
'Cristata'	'Hoseri'	'Sherwood Moss'
'Cristata Aurea'	'Hoveyi'	'Sphaerica'
'Danica'	'Intermedia'	'Sunkist'
'Dirigo Dwarf'	'Leptoclada'	'Tiny Tim'
'Dumosa'	'Little Champion'	'Umbraculifera'
'Ellwangeriana	'Little Gem'	'Van der Bom'
Aurea'	'Lombart's Dwarf'	'Wareana Globosa'
'Ericoides'	'Lutea Nana'	'Watereri'
'Filiformis'	'Meinekes Zwerg'	'Woodwardii'
'Froebellii'	'Nana'	
'Globosa'	'Ohlendorfii'	

Rounded

'Aurea'	'Ellwangeriana Aurea'	'Hoveyi'
'Boothii'	'Filiformis'	'Hudsonica'
'Caespitosa'	'Globosa'	'Little Gem'
'Canadian Green'	'Globosa Aurea'	'Little Giant'
'Conica Densa'	'Golden Globe'	'Lombart's Dwarf'
'Cristata'	'Green Midget'	'Meckii'
'Cristata Aurea'	'Hetz Midget'	'Meinekes Zwerg'
'Danica'	'Hoseri'	'Pumila Sudworthii'

'Rheindiana'   'Tiny Tim'   'Wagneri'
'Rheingold'   'Umbraculifera'   'Wareana Globosa'
'Sphaerica'   'Van der Bom'   'Woodwardii'

Yellow

'Aurea'   'George Peabody'   'Pumila Sudworthii'
'Aurea Americana'   'George Washington'   'Rheingold'
'Aureovariegata'   'Globosa Aurea'   'Rheingold Beattie'
'Aurescens'   'Golden'   'Riversii'
'Burrowii'   'Golden Globe'   'Semperaurea'
'Cloth of Gold'   'Hetz Midget   'Sudworthii'
'Cristata Aurea'   Variegated'   'Sunkist'
'Dorsett Yellow'   'Holmstrup Yellow'   'Vervaenaena'
'Douglasii Aurea'   'Lutea'   'Wareana Lutescens'
'Ellwangeriana Aurea'   'Lutea Nana'   'Watnong Gold'
'Europe Gold'   'Lutescens'   'Waxen'

White

'Alba'   'Elegantissima'   'Variegata'
'Beaufort'   'Meinekes Zwerg'   'Wansdyke Silver'
'Columbia'   'Sherwood Frost'

2. *T. plicata* J. Donn ex D. Don.   Western red-cedar. Giant arborvitae.

A native of northwestern North America that makes an excellent ornamental as well as fine telephone poles and cedar shakes. Leaves pointed, with triangular white markings on lower surface. Cones 1.3 cm (½ in) long, with 8–10 scales. Seeds winged and notched at the apex, usually 3 per scale.

'Atrovirens' Leaves dark green.

'Aurea' Leaves yellow.

'Aureovariegata' Leaves with yellow patches.

'Aurescens' Leaves at tips of young sprays yellow or pale green.

'Canadian Gold' Growth dwarf, conical; leaves yellow.

'Collyers Gold' Similar to 'Stoneham Gold'.

'Cuprea' Growth dwarf, conical, dense; twig tips nodding.

'Dura' Branches crowded; sprays long.

'Elegantissima'

'Euchlora'   Hardy; sprays many, long.

'Excelsa'   Growth columnar.

'Fastigiata'   Growth narrowly columnar.

'Gracilis'   Growth slow, conical; leaves smaller.

'Gracilis Aurea'   Growth dwarf; foliage lacelike with yellow.

'Green Sport'   Growth rapid, conical.

'Green Survival'   Growth conical; leaves remaining green.

'Hillieri'   Growth dwarf; twigs irregular in length.

'Hogan'   Growth tightly compact.

'Pendula'   Branches drooping.

'Pumila'   Growth dwarf, bun-shaped, low.

'Rogersii'   Growth dwarf, conical, dense; twigs crowded; leaves small, yellow.

'Rogersii Aurea'   Growth dwarf, conical to rounded; leaves small, yellow.

'Semperaurescens'   Young twigs and leaves with yellow tinge.

'Stoneham Gold'   Growth dwarf, rounded; foliage green in the center of the plant, yellow to the outside.

'Sunburst'   Foliage yellow.

'Sunshine'   Foliage yellow.

'Variegata'   Foliage patched with yellow.

'Virescens'   Foliage green.

'Zebrina'   Growth conical; leaves striped pale yellow or white.

3.   *T. standishii* (Gord.) Carr.   Japanese arborvitae. Standish arborvitae.

Handsome tree, native to Japan. Leaves on lateral twigs blunt, with white markings on lower surface. Cones 9–13 mm ($\frac{1}{3}$–$\frac{1}{2}$ in) long. Cone scales 10–12. Seeds 3 per scale, winged and not notched.

Thujopsis dolobrata

Thujopsis (L.f.) Sieb. & Zucc. Hiba-arborvitae. False-arborvitae. Broad-leaved-arborvitae.

This genus from Japan has a single species. It differs from *Thuja* in having broad twigs with axe-shaped lateral leaves and flat-topped cones with five seeds per wedge-shaped cone scale. The leaves are usually larger than the leaves of *Thuja*.

1. *T. dolobrata* (L.f.) Sieb. & Zucc. Hiba-arborvitae. False-arborvitae.

Native of Japan, where it grows to be a tree, but as a cultivated plant is most often a shrub. Not very hardy in the Northeast. Leaves with prominent white markings on lower surface. Cones 12–15 mm (½– ⅔ in) long.

'Nana' Growth dwarf; twigs straggly.
'Variegata' Twig tips white.

175

Torreya nucifera

Torreya Arn.   Torreya.   Nutmeg.

This eastern Asian and North American genus of seven species resembles *Taxus* and *Cephalotaxus,* but it is very unlikely to be found in northeastern North America outside botanical gardens or arboreta. Only two species are hardy as far north as New York City. The genus is distinguished from *Taxus* by its opposite branches and purple female reproductive structures and from *Cephalotaxus* by the lack of a stalk on female reproductive structures and the lack of a prominent midrib on the leaves.

1. Leaves 4–10 cm (1½–4 in) long . . . . *T. californica*
California torreya
1. Leaves 2.0–3.3 cm (¾–1¼ in) long . . . . *T. nucifera*
Torreya

1. *T. californica* **Torr.**   **California torreya.**   **California-nutmeg.**

> Shrub native to California. Probably hardy only as far north as New York City area. Leaves 4–10 cm (1½–4 in) long. Reproductive structure 2–3 cm (¾–1¼ in) wide and 2.5–4.0 cm (1–1½ in) long.

2. *T. nucifera* **(L.) Sieb. & Zucc.**   **Torreya.**   **Japanese torreya.**

> Tree native to Japan. Hardier and found in cultivation more frequently than *T. californica*. Leaves 2.0–3.3 cm (¾–1¼ in) long. Reproductive structure 1.2–2.0 cm (½–¾ in) wide and 2.0–2.5 cm (¾–1 in) long.

> 'Aurea Variegata'   Foliage patched yellow.
> 'Prostrata'   Growth dwarf, low.

Tsuga canadensis

Tsuga Carr. Hemlock.

146. *T. caroliniana*

147. *T. diversifolia*

148. *T. chinensis*

This genus of North America and Asia includes 18 species and is abundant in northeastern North America, both as a native tree and as an ornamental planting. It is characterized by short, flat, blunt leaves of different sizes on the same branch. The branches are flexible, giving the trees a soft, pliant, graceful appearance, with a wavy, drooping leading shoot for most of the year. The cones are small and hang from the ends of the twigs.

1. Leaves with white longitudinal bands on both
   surfaces . . . . *T. mertensiana*
   Mountain hemlock
1. Leaves with white longitudinal bands on lower surface
   only.
   2. Edges of leaves smooth and unbroken (Fig. 146).
      3. Twigs smooth, lacking hairs . . . . *T. sieboldii*
         Southern Japanese hemlock
      3. Twigs hairy.
         4. Twigs hairy throughout (Fig. 147) . . . . *T. diversifolia*
            Northern Japanese hemlock
         4. Twigs hairy only in the grooves (Fig. 148).

178

149. *T. chinensis*

150. *T. caroliniana*

151. *T. canadensis*

5. Leaves with inconspicuous white longitudinal bands on lower surface and notched at the tip (Fig. 149) . . . . *T. chinensis*
Chinese hemlock

5. Leaves with conspicuous white longitudinal bands on lower surface and rounded or obscurely notched at the tip (Fig. 150) . . . . *T. caroliniana*
Carolina hemlock

2. Edges of leaves minutely toothed (Fig. 151).

6. Leaves on first-year shoots tapering, wider at the base than at the tip, mostly in several planes about the twig; common . . . . *T. canadensis** +
Canadian hemlock

6. Leaves on first-year shoots not tapering, the tip about as wide as the base, mostly in one plane; rare . . . . *T. heterophylla*
Western hemlock

* + 1.   *T. canadensis* L.   Canadian hemlock.

A native timber tree of the eastern North American forests that has been planted in abundance. A vast number of cultivars available for all ornamental purposes. Leaves with minute teeth toward the tip (visible with hand lens), 7–17 mm (¼–⅔ in) long. Cones 1.3–2.0 cm (½–¾ in) long.

'Abbott's Dwarf'   Growth slow, conical.

'Abbott's Fountain'

'Abbott's Pygmy'   Growth dwarf.

'Abbott Weeping'   Growth dwarf, low; some leaves twisted.

'Albopicta'   Growth dwarf, rounded; leaves at branch tips white.

'Albospica'   Growth compact; leaves at twig tips white.

'Andrews'   Growth dwarf.

'Angustifolia'   Bushy tree; branches in clusters; leaves about 10 times longer than wide.

'Armistice'   Growth slow, flat-topped; branches irregular.

'Ashfield Weeper'   Branches drooping.

'Atrovirens'   Leaves very dark green.

'Aurea'   Branches stiff, coarse; leaves yellow in first year.

'Aurea Compacta'   Growth dwarf, compact; foliage yellow.

'Bacon Cristate'   Growth dwarf; leaves dark green.

'Bagatelle'   Growth compact.

'Baldwin Dwarf Pyramid'   Growth slow, conical.

'Barrie Bergman'   Growth slow, flat-topped.

'Beaujean'   Growth low; branches thick, directed back toward
   stem and sideways.
'Beehive'   Growth dwarf, rounded.
'Bennett'   Growth low, spreading; branches drooping.
'Bennett Golden'   Foliage yellow.
'Bennett's Minima'   Same as 'Bennett'.
'Bergman's Cascade'   Growth dwarf.
'Bergman's Frosty'   Same as 'Frosty'.
'Bergman's Gem'   Growth dwarf, rounded.
'Betty Rose'   Shrub; growth dwarf, compact; spring and summer foliage
   white.
'Bonnie Bergman'   Growth dwarf.
'Boulevard'   Growth conical; some branches long; leaves
   crowded, long.
'Bradshaw'   Growth conical, compact.
'Brandley'   Growth slow, conical, compact; twigs narrow, heavy.
'Brevifolia'   Leaves short.
'Bristol'   Growth rounded or broad-topped, bushy, multistemmed.
'Broad Globe'
'Brookline'   Growth dwarf; branches drooping.
'Broughton'   Growth slow, conical, irregular; leaves crowded, long.
'Buck Estate'
'Callicoon'   Growth twice as wide as high; branches arching.
'Calvert'   Growth slow; foliage dense, tufted.
'Cappy's Choice'   Growth compact, low; leaves tinged with yellow.
'Cinnamomea'   Growth dwarf, wider than high; buds light brown; twigs
   with brown hairs.
'Classic'   Growth wide; branches twisted.
'Cloud Prune'   Growth dwarf; leaves irregularly spaced.
'Coffin'   Growth slow, compact; foliage clumped at twig tips.
'Cole'   Same as 'Cole's Prostrate'.
'Cole's Prostrate'   Growth low to the ground; branches drooping.
'Columnaris'   Growth columnar.
'Compacta'   Growth dwarf, conical.
'Compacta Aurea'   Same as 'Everitt Golden'.
'Contorted'   Twigs with twisted tips.
'Coplen'   Growth conical, compact.
'Copley's Pyramidal'
'Corbit'   Growth dwarf, dense.
'Creamey'   Growth slow, foliage patched cream color.

'Curly'   Growth dwarf; leaves crowded, curved around the stems.

'Curtis Ideal'   Growth dwarf, conical but wider than high.

'Curtis Spreader'   Growth dwarf, spreading, low, dense.

'Cushion'   Growth bun-shaped.

'David Verkade'   Branches drooping.

'Dawsoniana'   Growth slow, compact; leaves dark green.

'Densifolia'   Leaves close together.

'Densiforma'

'Detmer's Weeper'   Branches drooping.

'Diversifolia'   Branches twisted at the ends.

'Doc's Choice'   Growth slow, conical; foliage dark green.

'Doran'   Growth dwarf, compact, appearing sheared; branches crowded, tips drooping.

'Dover'   Growth compact; leaves large.

'Drake'   Growth rounded.

'Dr. Hornbeck'   Same as 'Hornbeck'.

'Droop Tip'   Growth dwarf; branch tips drooping.

'Dwarf Whitetip'   Growth conical; young leaves white.

'Elm City'   Growth dwarf; branches drooping somewhat.

'Essex'   Similar to 'Minuta'.

'Everitt Dense Leaf'   Similar to 'Hussii'.

'Everitt Golden'   Growth dwarf, dense; foliage yellow.

'Fantana'   Growth spreading, wider than high.

'Far Country'   Growth dwarf, multistemmed.

'Fastigiata'   Growth columnar.

'Feasterville'   Growth conical; leaves small, pointing forward.

'Fremdii'   Growth slow, conical; leaves dark green.

'Frosty'   Foliage white if kept in shade.

'Gable Weeping'   Growth compact; branches drooping.

'Geneva'   Growth dwarf, upright.

'Gentsch'   Growth dwarf, mound-shaped; leaves with white tips.

'Gentsch Dwarf'   Growth dwarf, rounded.

'Gentsch Dwarf Globe'   Same as 'Gentsch Globe'.

'Gentsch Globe'   Growth dwarf, rounded, compact.

'Gentsch Snowflake'   Same as 'Snowflake'.

'Gentsch Variegated'   Growth dwarf, rounded, dense; young shoots radiating from the center of the plant, exposed, reddish; some leaves white.

'Gentsch White'   Growth dwarf, mound-shaped; leaves at twig tips white.

'Gentsch White Tip'   Growth slow, rounded; foliage patched white.

'Globosa'   Growth dwarf, rounded, compact.

'Golden Splendor'   Growth upright; foliage yellow.

'Gracilis'   Growth dwarf; branches drooping at tips; leaves very small.

'Gracilis Nana'   Branches long; leaves rigid, short.

'Great Lakes'   Growth conical, dense.

'Green Cascade'   Growth dwarf, dense; branches drooping.

'Greenspray'   Growth dwarf, mounded.

'Greenwood Lake'   Growth slow; branches crowded, irregular.

'Guldemond's Dwarf'   Growth conical, irregular, dense.

'Hahn'   Growth dwarf, spreading.

'Hancock'   Growth dwarf; twigs crowded.

'Harmon'   Growth columnar, dense, becoming wide and loose.

'Heckman'

'Helene Bergman'   Growth dwarf, dense.

'Heli'   Growth dwarf; leaves wide.

'Henry Hohman'   Growth dwarf, conical, dense.

'Hicks'   Similar to 'Atrovirens'.

'Hiti'   Growth conical, compact when young.

'Horbeck'   Growth dwarf, conical, wider than high; leaves small.

'Horsford'   Growth dwarf, irregular, dense.

'Horsford Contorted'   Growth slow; branches twisted.

'Horsford Dwarf'   Growth dwarf, congested; twigs curved down; leaves short, crowded.

'Horton'   Branches drooping.

'Howe'   Growth conical, compact.

'Huffman's Compact'

'Hunnewell'   Branches drooping.

'Hussii'   Growth dwarf, conical; branches short and crowded.

'Ideal'

'Imperial'   Growth compact; branch tips white.

'Innisfree'   Growth rounded.

'Jacqueline Verkade'   Growth conical, dense; leaves small.

'Jan Verkade'   Growth rapid, spreading.

'Jeddeloh'   Growth dwarf, spreading, then mound-shaped; branches drooping.

'Jenkinsii'   Growth rapid; leaves pressed close to twig.

'Jennings Yewlike'   Growth rapid; leaves irregular.

'Jervis'   Growth dwarf, compact; branches crowded.

'John Swartley'   Growth slow, spreading.

'Juniperoides'   Same as 'Feasterville'.

'Kathryn Verkade'   Growth dwarf, spreading, flat-topped; leaves small.

'Kelseyi'   Same as 'Kelsey's Weeping'.

'Kelsey's Weeping'   Branches drooping.

'Kingston Hollow'   Growth dwarf; leaves short.

'Kingsville'   Growth columnar.

'Kingsville Spreader'   Growth dwarf, rounded; branches drooping.

'LaBar Gem'   Growth conical, compact, stiff.

'LaBar White Tip'   Foliage white in summer.

'Latifolia'   Branches long, horizontal, then turned down; leaves crowded.

'Laurie'   Growth slow, rounded.

'Lewis'   Growth dwarf, conical; branches thick; leaves tightly pressed to the branches.

'Lincoln White Tip'

'Little Joe'   Growth dwarf, rounded; branches crowded.

'Loudon Dwarf'

'Lustgarten Creeping'   Branches spreading, drooping.

'Macrophylla'   Growth rapid; leaves larger.

'Mansfield'   Growth rounded; branches crowded, close to the ground, tangled with each other.

'Matthews'   Growth rounded, compact when young.

'Meyers'   Growth conical, compact.

'Microphylla'   Leaves short.

'Milfordensis'   Growth dwarf, rounded; branches crowded; twigs slender; leaves short.

'Minima'   Growth dwarf; branches spreading.

'Minuta'   Growth dwarf, compact; branches crowded, irregularly spaced; leaves short.

'Moll'   Growth conical, compact.

'Moon's Columnar'   Growth columnar, dense.

'Muttontown'   Same as 'Armistice'.

'Nana'   Growth dwarf, spreading, wider than high.

'Nana Gracilis'   Same as 'Gracilis'.

'Narragansett'   Branches spreading.

'Nearing'   Similar to 'Jervis'.

'Oldenburg'   Growth dwarf, rounded; depressed area in middle of plant in early years; twigs very short.

'Outpost'   Growth rapid; twigs drooping; leaves close together.

'Palomino'   Growth dwarf, rounded, compact.

'Parvifolia'   Twigs numerous; leaves very small.

'Pendula'   Growth rounded, wider than high; branches and twigs drooping.

'Pendula Argentea'   Branches drooping; leaves white near twig tips.

'Plainview'

'Pomfret'   Growth conical, dense.

'Popeleski'   Growth dwarf, conical, compact.

'Pumila'   Growth dwarf, conical; branch tips sharply drooping.

'Pygmaea'   Same as 'Abbott's Pygmy'.

'Raraflora Snowflake'

'Redding'   Foliage dense.

'Rock Creek'   Growth rounded, compact.

'Rockland'   Foliage dense.

'Rockport'   Growth dwarf; leaves small.

'Ruggs'   Leaves brown at tips.

'Ruggs Washington Dwarf'   Growth dwarf, rounded.

'Randy Small Leaf'   Leaves small, widely spaced, pressed close to the twigs.

'Salicifolia'   Growth compact; leaves small.

'Sargentii'   Branches drooping.

'Schramm'   Growth columnar.

'Sherwood'   Growth spreading.

'Sherwood Compact'   Growth low, spreading, rounded; branches somewhat twisted.

'Silver Tip'   Leaves at twig tips white.

'Silvery Gold'   Similar to 'Everitt Golden'.

'Slenderella'   Growth rapid; leaves small.

'Snowflake'   Growth rounded; foliage white in summer.

'Sparsifolia'   Growth compact but foliage lacy.

'Starker'   Growth dwarf, much wider than high; branches drooping.

'Stewartii'   Same as 'Stewart's Gem'.

'Stewart's Dwarf'   Growth dwarf, mound-shaped, low.

'Stewart's Dwarf Globe'   Growth dwarf, rounded, compact.

'Stewart's Gem'   Growth dwarf, compact.

'Stockman's Compact'   Growth slow.

'Stockman's Dwarf'   Growth slow; leaves short, thick and close together.

'Stranger'   Growth slow, compact, as wide as high.

'Taxifolia'   Leaves wider, longer, spreading at all angles from the twig.

'Thurlow' Growth rapid, compact; leaves small.

'T. L. Edwards'

'Towson' Growth compact; leaves large.

'Unique' Growth slow, rounded, dense.

'Upper Bank' Growth conical, slow.

'Valentine' Growth slow; branches drooping.

'Van Dyne'

'Verkade Petite' Growth dwarf, rounded, dense.

'Verkade's Recurved' Growth dwarf; leaves curved.

'Vermeulen's Pyramid' Growth columnar; leaves close together.

'Von Helms' Growth dwarf, dense; leaves long and wide; cone tips blunt.

'Von Helms Dwarf' Same as 'Von Helms'.

'Warner's Globe' Growth dwarf, rounded; leaves light green.

'Watnong Star' Shrub; growth dwarf, rounded, dense; new foliage white.

'Waverly' Foliage dense.

'Wellesleyana' Growth conical, low; branches drooping.

'West Coast Creeper' Growth spreading, drooping.

'Westonigra' Growth compact, upright, wide; branches drooping; leaves dark green.

'Wheelerville' Growth slow, conical; leaves close together.

'White Tip'

'Wilton' Twigs drooping.

'Wilton Globe' Growth rounded, multistemmed.

'Wodenethe' Branches drooping.

'Yellow Tip'

'Youngcone' Growth upright; branches drooping; cones numerous, appearing early.

Tsuga canadensis Cultivar Character Groupings

Dwarf

'Abbott's Pygmy'	'Bennett's Minima'	'Cloud Prune'
'Abbott Weeping'	'Bergman's Cascade'	'Compacta'
'Albopicta'	'Bergman's Gem'	'Compacta Aurea'
'Andrews'	'Betty Rose'	'Corbit'
'Aurea Compacta'	'Bonnie Bergman'	'Curly'
'Bacon Cristate'	'Brookline'	'Curtis Ideal'
'Beehive'	'Cinnamomea'	'Curtis Spreader'

'Doran'
'Dr. Hornbeck'
'Droop Tip'
'Elm City'
'Essex'
'Everitt Dense Leaf'
'Everitt Golden'
'Far Country'
'Geneva'
'Gentsch'
'Gentsch Dwarf'
'Gentsch Dwarf Globe'
'Gentsch Globe'
'Gentsch Variegated'
'Gentsch White'
'Globosa'
'Gracilis'
'Green Cascade'
'Greenspray'
'Hahn'
'Hancock'

'Helene Bergman'
'Heli'
'Henry Hohman'
'Hornbeck'
'Horsford'
'Horsford Dwarf'
'Hussii'
'Jeddeloh'
'Jervis'
'Kathryn Verkade'
'Kingston Hollow'
'Kingsville Spreader'
'Lewis'
'Little Joe'
'Milfordensis'
'Minima'
'Minuta'
'Nana'
'Nearing'
'Oldenburg'
'Palomino'

'Popeleski'
'Pumila'
'Pygmaea'
'Rockport'
'Ruggs Washington
  Dwarf'
'Silvery Gold'
'Starker'
'Stewart's Dwarf'
'Stewart's Dwarf Globe'
'Stewart's Gem'
'Verkade Petite'
'Verkade's Recurved'
'Von Helms'
'Von Helms Dwarf'
'Warner's Globe'
'Watnong Star'

Slow

'Abbott's Dwarf'
'Armistice'
'Baldwin Dwarf Pyramid'
'Barrie Bergman'
'Brandley'
'Broughton'
'Coffin'
'Creamey'

'Dawsoniana'
'Doc's Choice'
'Fremdii'
'Gentsch White Tip'
'Greenwood Lake'
'Horsford Contorted'
'John Swartley'
'Laurie'

'Muttontown'
'Stockman's Compact'
'Stockman's Dwarf'
'Stranger'
'Unique'
'Upper Bank'
'Valentine'
'Wheelerville'

Rounded

'Albopicta'
'Beehive'
'Bergman's Gem'
'Bristol'
'Cushion'
'Drake'
'Gentsch'
'Gentsch Dwarf'
'Gentsch Dwarf Globe'
'Gentsch Globe'

'Gentsch Snowflake'
'Gentsch Variegated'
'Gentsch White'
'Gentsch White Tip'
'Globosa'
'Greenspray'
'Innisfree'
'Jeddeloh'
'Kingsville Spreader'
'Laurie'

'Little Joe'
'Mansfield'
'Matthews'
'Milfordensis'
'Oldenburg'
'Palomino'
'Pendula'
'Rock Creek'
'Ruggs Washington Dwarf'
'Sherwood Compact'

'Snowflake'   'Unique'   'Warner's Globe'
'Stewart's Dwarf'   'Verkade Petite'   'Wilton Globe'
'Stewart's Dwarf Globe'

Yellow

'Aurea'   'Cappy's Choice'   'Golden Splendor'
'Aurea Compacta'   'Compacta Aurea'   'Silvery Gold'
'Bennett's Golden'   'Everitt Golden'

White

'Albopicta'   'Gentsch'   'LaBar White Tip'
'Albospica'   'Gentsch   'Pendula Argentea'
'Bergman's Frosty'   Snowflake'   'Silver Tip'
'Betty Rose'   'Gentsch   'Snowflake'
'Creamey'   Variegated'   'Watnong Star'
'Dwarf Whitetip'   'Gentsch White'   'White Tip'
'Frosty'   'Gentsch White Tip'

2. *T. caroliniana* Engelm.   Carolina hemlock.

Native to the southeastern United States. Grows well in cultivation, although it is infrequently planted. Distinguished from *T. canadensis* by its more compact growth habit, although its leaves are spaced farther apart on the branches. Leaves 1–2 cm (⅓–¾ in) long. Cones 2.5–3.8 cm (1–1½ in) long.

'Adams Weeping'   Branches drooping; leaves irregularly spaced.
'Arnold Pyramid'   Growth conical, very dense.
'Ashford'   Growth dense; leaves shorter.
'Compacta'   Growth rounded, dense.
'Elizabeth Swartley'   Branches drooping.
'LaBar Bushy'   Growth rounded, loose, multistemmed.
'LaBar Weeping'   Growth slow, dense; branches drooping.
'Warner Weeping'   Branches drooping.

3. *T. chinensis* (Franch.) Pritz.   Chinese hemlock.

Native to China. Not cultivated to any great extent, partly because of its slow growth. Leaves 1.3–2.5 cm (½–1 in) long. Cones 1.5–2.5 cm (⅗–1 in) long.

4. *T. diversifolia* (**Maxim.**) **M. T. Mast.   Northern Japanese hemlock.**

Native to Japan. Grows well in cultivation. Infrequently planted. Branches closer together on the trunk than those of *T. sieboldii*. Leaves 4–14 mm (⅕–⅗ in) long. Cones 1.3–2.0 cm (½–¾ in) long.

'Gotelli'   Growth dwarf, dense; leaves shorter.
'Manifold'   Growth slow, compact.
'Medford Lake'   Growth slow, irregular.
'Thompson'

5. *T. heterophylla* (**Raf.**) **Sarg.   Western hemlock.**

Native to northwestern North America. Rarely cultivated in the Northeast. Leaves 0.5–2.0 cm (½–¾ in) long. Cones 2.0–2.5 cm (¾–1 in) long.

'Conica'   Growth conical.
'Dumosa'   Growth dwarf, bushy; twigs short, stiff.
'Epstein Dwarf'   Growth dwarf.
'Greenmantle'   Branches drooping.
'Iron Springs'   Growth dwarf, wide, spreading; leaves shorter.
'Laursen's Column'   Branches drooping; foliage dense.
'Sixes River'

6. *T. mertensiana* (**Bong.**) **Carr.   Mountain hemlock.**

Hemlock of the mountains of western North America. A small tree when planted in the northeastern part of the continent, where it does not grow well. Leaves 1.3–2.5 cm (½–1 in) long. Cones 5–8 cm (2–3 in) long.

'Argentea'   Growth weak; leaves silver.
'Argenteovariegata'   Leaves on twig tips white.
'Blue Cloud'   Growth compact; leaves blue.
'Blue Star'   Leaves blue.
'Cascade'   Growth dwarf, compact; leaves shorter, closer together.

'Columnaris' Growth columnar, compact.

'Elizabeth' Growth dwarf, spreading.

'Emile's Select' Growth slow, conical.

'Glauca' Growth slow, compact; leaves blue-green.

'Mount Arrowsmith' Growth slow, conical, compact, irregular; leaves small.

'Mount Hood' Growth dwarf, dense, irregular.

'Murthly Castle' Branches drooping.

'Quartz Mountain' Growth slow, conical.

'Sherwood Compact' Same as 'Mount Hood'.

'Van's Prostrate'

7. *T. sieboldii* Carr. Southern Japanese hemlock.

Native to Japan. Infrequently cultivated in northeastern North America. Leaves 0.7–2.0 cm (¼–¾ in) long. Cones 2.5–3.3 cm (1–1¼ in) long.

'National' Growth rapid, irregular.

Conifer Cultivar Character Groupings

The groupings in this appendix include most, but not all, of the cultivars listed in the species descriptions. In addition, cultivars that have a conical shape, green foliage, *and* a normal growth rate—that is, typical conifer characters—are not included.

Growth Rate Dwarf

Foliage All or Partly Yellow

*Chamaecyparis
  lawsoniana*

'Aurea Densa'
'Lutea Nana'
'Minima Aurea'

Chamaecyparis obtusa

'Aurea Nana'
'Fernspray Gold'
'Kojolkohiba'
'Nana Aurea'
'Nana Lutea'
'Opaal'
'Tetragona Aurea'
'Verdoni'
'Yellowtip'

*Chamaecyparis
  pisifera*

'Aurea Nana'

'Compressa Aurea'
'Filifera Aurea Nana'
'Filifera
  Aureovariegata'
'Flavescens'
'Gold Dust'
'Golden Mop'
'Juniperoides Aurea'
'Lutea'
'Mikko'
'Minima Aurea'
'Minima Variegata'
'Mops'
'Nana Aurea'
'Nana Aureovariegata'
'Plumosa Nana Aurea'
'Plumosa Rogersii'
'Snow'
'Strathmore'
'Sulphurea Nana'
'Winter Gold'

Juniperus communis

'Ellis'
'Nana Aurea'

Picea abies

'Diffusa'
'Ohlendorfii'

Picea orientalis

'Aurea Compacta'
'Compacta Aurea'
'Early Gold'
'Skylands'

Pinus mugo

'Aurea'
'Winter Gold'

Pinus sylvestris

'Aureopicta'
'Gold Coin'

'Greg's Variegated'
'Moseri'

Platycladus orientalis
'Hillieri'
'Semperaurescens'

Taxus baccata
'Elegantissima'
'Pumila Aurea'
'Repandens Aurea'
'Standishii'

Taxus cuspidata
'Fastigiata'

Thuja occidentalis
'Cristata Aurea'
'Ellwangeriana Aurea'
'Globosa Aurea'
'Golden Globe'
'Hetz Midget
Variegated'
'Lutea Nana'
'Pumila Sudworthii'
'Rheingold'
'Sunkist'

Thuja plicata
'Canadian Gold'
'Gracilis Aurea'

'Rogersii'
'Rogersii Aurea'
'Stoneham Gold'

Tsuga canadensis
'Aurea Compacta'
'Compacta Aurea'
'Everitt Golden'
'Silvery Gold'

Foliage All or Partly White

Cedrus deodara
'Nivea'
'Silver Mist'
'Snow Sprite'
'White Imp'

*Chamaecyparis
lawsoniana*
'Croftway'
'Ellwood's White'
'Fleckalwood'
'Minima'
'Monumentalis
Glauca'
'Nana Albospica'

Chamaecyparis obtusa
'Nana Argentea'
'Snowkist'
'Tonia'
'White Tip'

*Chamaecyparis
pisifera*
'Compacta Variegata'
'Filifera
Argenteovariegata'
'Nana Albovariegata'
'Nana Variegata'
'Silver Lode'
'Snow'
'White Pygmy'

Cryptomeria japonica
'Albovariegata'
'Knaptonensis'
'Nana Albospica'
'Okina-sugi'

Juniperus sabina
'Variegata'

Pinus thunbergiana
'Tigrina'
'Tura-ku Kuromatsu'

Platycladus orientalis
'Pumila Argenta'
'Summer Cream'

Sequoia sempervirens
'Adpressa'
'Albospica'

Thuja occidentalis
'Meinekes Zwerg'

Tsuga canadensis
'Albopicta'
'Betty Rose'
'Dwarf Whitetip'
'Far Country'
'Gentsch'
'Gentsch Variegated'
'Gentsch White'
'Watnong Star'

Foliage Green (Species Normal)

Abies alba
'Compacta'

'Elegans'
'Microphylla'

Abies balsamea
'Hudsonia'

Abies cephalonica
  'Myer's Dwarf'

Abies concolor
  'Archer's Dwarf'
  'Compacta'
  'Gable's Weeping'
  'Globosa'
  'Green Globe'
  'Masonic Broom'
  'Vineola Dwarf'

Abies grandis
  'Nana'

Abies homolepis
  'Scottae'
  'Tomonii'

Abies koreana
  'Compact Dwarf'
  'Nisbet'

Abies lasiocarpa
  'Beissneri'
  'Compacta'
  'Conica'
  'DuFlon'
  'Roger Watson'

Abies nordmanniana
  'Horizontalis'
  'Nana'

Abies pinsapo
  'Clarke'
  'Horstmann'
  'Nana'

Abies procera
  'Prostrata'

Calocedrus decurrens
  'Intricata'

Cedrus deodara
  'Descano Dwarf'
  'Hesse'

  'Nana'
  'Pygmea'

Cedrus libani
  'Comte de Dijon'
  'Conica Nana'
  'Nana'
  'Nana Pyramidata'
  'Sargentii'

Cephalotaxus fortunei
  'Concolor'

*Cephalotaxus
  harringtonia*
  'Gnome'

*Chamaecyparis
  lawsoniana*
  'Chilworth Silver'
  'Compacta'
  'Dwarf Blue'
  'Ellwood's Pygmy'
  'Erecta Aurea'
  'Ericoides'
  'Filiformis Compacta'
  'Fletcheri Nana'
  'Fletcher's Compact'
  'Forsteckensis'
  'Gimbornii'
  'Globosa'
  'Gnome'
  'Grandis'
  'Knowefeldensis'
  'Little Spire'
  'Minima Glauca'
  'Nana'
  'Nana Argentea'
  'Nana Glauca'
  'Nana Rogersii'
  'Nidiformis'
  'Pygmaea'
  'Pygmaea Argentea'
  'Rogersii'
  'Silver Thread'
  'Tharandtensis'
  'Tharandtensis Caesia'
  'Wansdyke Miniature'
  'Wisselii Nana'

Chamaecyparis obtusa
  'Barkenny'
  'Bassett'
  'Bess'
  'Buttonball'
  'Caespitosa'
  'Chabo-yadori'
  'Chilworth'
  'Chimohiba'
  'Compacta'
  'Coralliformis'
  'Coralliformis Nana'
  'Densa'
  'Ericoides'
  'Flabelliformis'
  'Gracilis Nana'
  'Graciosa'
  'Hage'
  'Intermedia'
  'Juniperoides'
  'Juniperoides
   Compacta'
  'Kaanamihiba'
  'Kamaeni Hiba'
  'Kamakura Hiba'
  'Kosteri'
  'Kosteri Nana'
  'Laxa'
  'Minima'
  'Nana'
  'Nana Compacta'
  'Nana Contorta'
  'Nana Densa'
  'Nana Gracilis'
  'Nana Kosteri'
  'Nana Prostrata'
  'Nana Pyramidalis'
  'Nana Repens'
  'Pygmaea'
  'Pygmaea Aurescens'
  'Pygmaea Densa'
  'Reis'
  'Reis Dwarf'
  'Repens'
  'Rigid Dwarf'
  'Spiralis'
  'Split Rock'
  'Stoneham'

'Suirova-hiba'
'Tetragona'
'Tetragona Minima'
'Tiny Tot'
'Torulosa Nana'
'Tsatsumi'

*Chamaecyparis
pisifera*
'Compacta'
'Compacta Nana'
'Dwarf Blue'
'Ericoides'
'Filifera Nana'
'Glauca Compacta
Nana'
'Hime-savara'
'Minima'
'Nana'
'Nana Compacta'
'Parslori'
'Plumosa Aurea Nana'
'Plumosa Compressa'
'Plumosa Cristata'
'Plumosa
Juniperoides'
'Plumosa Nana
Variegata'
'Plumosa Pygmaea'
'Pygmy'
'Squarrosa Cristata'
'Squarrosa Intermedia'
'Squarrosa Nana'
'Squarrosa Pygmaea'
'Squarrosa Sieboldii'
'Tsukumi'

*Chamaecyparis
thyoides*
'Andelyensis Aurea'
'Andelyensis Conica'
'Andelyensis Nana'
'Conica'
'Ericoides'
'Little Jamie'
'Nana'
'Pygmaea'
'Rezek's Dwarf'
'Rubicon'

Cryptomeria japonica
'Araucarioides'
'Chabo-sugi'
'Compacta Nana'
'Eizan-sugi'
'Elegans Compacta'
'Elegans Nana'
'Enko-sugi'
'Fasciata'
'Globosa Nana'
'Kewensis'
'Kilmacurragh'
'Lobbii Nana'
'Mankichi-sugi'
'Mankitiana-sugi'
'Mejero-sugi'
'Monstrosa Nana'
'Nana'
'Osaka-tama-sugi'
'Pungens'
'Pygmaea'
'Rein's Dense Jade'
'Shishi-gashirad'
'Tansu'
'Vilmoriniana'
'Yatsubusa'
'Yatsubusa-sugi'
'Yatsufusa'

Cupressus arizonica
'Compacta'
'Crowborough'
'Gareei'

Juniperus chinensis
'Rockery Gem'
'San Jose'
'Sea Spray'
'Shoosmith'
'Titlis'

Juniperus communis
'Bakony'
'Berkshire'
'Compressa'
'Depressed Star'
'Derrynana'
'Dumosa'
'Echiniformis'

'Effusa'
'Gew Graze'
'Gold Beach'
'Minima'
'Nana'
'Repanda'
'Soapstone'
'Suecica Nana'
'Vase'
'Windsor Gem'
'Zeal'

Juniperus horizontalis
'Admirabilis'
'Adpressa'
'Blue Rug'
'Blue Wilton'
'Emerson'
'Filicina Minima'
'Glomerata'
'Wiltonii'

Juniperus procumbens
'Nana'
'Nana Glauca'
'Santarosa'

Juniperus sabina
'Arcadia'
'Blue Forest'
'Rockery Gem'
'Thomsen'

Juniperus scopulorum
'Gareei'
'Hillburn's Silver
Globe'
'Repens'
'Silver King'
'Tabletop'

Juniperus squamata
'Blue Star'
'Prostrata'
'Pygmaea'

Juniperus virginiana
'Humilis'
'Kobold'

'Nana'
'Nana Compacta'
'Pendula Nana'
'Pumila'

Larix decidua
'Repens'

Larix kaempferi
'Minor'
'Varley'
'Wehlen'

Picea abies
'Barnes'
'Beissneri'
'Bennett's Miniature'
'Capitata'
'Compact Asselyn'
'Conica'
'Costickii'
'Crippsii'
'Decumbens'
'Dellensis'
'Doversii Pendula'
'Dumosa'
'Echiniformis'
'Elegans'
'Formanek'
'Gracilis'
'Gregoryana'
'Gregoryana Parsonsii'
'Gregoryana Veitchii'
'Hillside Dwarf'
'Holmstrup'
'Hornibrookii'
'Humilis'
'Hystrix'
'Kalmthout'
'Kamon'
'Kluis'
'Knaptonensis'
'Little Gem'
'Little Joe'
'Mariae Orffi'
'Maxwellii'
'Microphylla'
'Microsperma'
'Minutifolia'

'Nana'
'Nana Compacta'
'Nidiformis'
'Pachyphylla'
'Parsonsii'
'Procumbens'
'Pseudoprostrata'
'Pumila'
'Pumila Glauca'
'Pumila Nigra'
'Pumilio'
'Pygmaea'
'Pyramidalis Gracilis'
'Repens'
'St. James'
'Sargentii'
'Tabuliformis'
'Veitchii'
'Wagneri'
'Wansdyke Miniature'
'Waugh'
'Wills Zwerg'
'Wilson'

Picea alcoquiana
'Howell's Dwarf'

Picea asperata
'Hunnewelliana'

Picea engelmannii
'Compacta'
'Microphylla'

Picea glauca
'Alberta Globe'
'Cecilia'
'Compacta'
'Conica'
'Cy's Wonder'
'Echiniformis'
'Gnome'
'Hillside'
'Laurin'
'Lilliput'
'Millstream Broom'
'Monstrosa Nana'
'Nana'
'Pixie'

'Sander's Blue'
'Tabuliformis'
'Tiny'

Picea jezoensis
'Nana'

Picea mariana
'Doumettii'
'Globosa'
'Empetroides'
'Nana'
'Procumbens'
'Pygmaea'
'Semiprostrata'

Picea omorika
'Berliners Weeping'
'Expansa'
'Frohnleiten'
'Gnom'
'Minima'
'Nana'
'Pimoko'

Picea orientalis
'Gracilis'
'Mount Vernon'
'Nana'
'Weeping Dwarf'

Picea pungens
'Blue Trinket'
'Egyptian Pyramid'
'Globosa'
'Gotelli's Broom'
'Hunnewelliana'
'Luusbarg'
'Moll'
'Montgomery'
'Mrs. Cessarini'
'Nana'
'Prostrate Blue Mist'
'St. Mary'
'St. Mary's Broom'

Picea sitchensis
'Compacta'
'Papoose'

'Strypemonde'
'Tenas'
'Upright Dwarf'

Pinus albicaulis
'Flinck'
'Nana'
'Nobles Dwarf'
'Number One Dwarf'

Pinus aristata
'Baldwin Dwarf'
'Cecilia'
'Sherwood Compact'

Pinus banksiana
'Baba'
'Uncle Fogy'

Pinus cembra
'Chalet'
'Globe'
'Jermyns'

Pinus contorta
'Minima'
'Spaan's Dwarf'

Pinus densiflora
'Alice Verkade'
'Heavy Bud'
'Jane Kluis'
'Umbraculifera'

Pinus echinata
'Clines Dwarf'

Pinus flexilis
'Nana'
'Witch's Broom'

Pinus koraiensis
'Dwarf'
'Winton'

Pinus leucodermis
'Compact Gem'
'Schmidtii'

Pinus mugo
'Allen's Seedling'
'Alpenglow'
'Compacta'
'Elfengren'
'Gnom'
'Green Candle'
'Hesse'
'Humpy'
'Kissen'
'Knapenburg'
'Kobold'
'Mops'
'Oregon Jade'
'Oregon Pixie'
'Sherwood Compact'
'Slavinii'
'Teeny'
'Trompenburg'
'Tyrol'

Pinus nigra
'Balcanica'
'Bujotii'
'Nana'

Pinus parviflora
'Adcock's Dwarf'
'Baldwin'
'Bergmanii'
'Glauca Compacta'
'Glauca Nana'
'Kokouse'
'Koraku'
'Nana'
'Nasu'
'Pygmy Yatsubusa'
'Setsu-gek-ka'
'Shiobara Yatsubusa'
'Venus'
'Yatsubusa'
'Yu-ho'

Pinus peuce
'Arnold Dwarf'
'Nana'

Pinus ponderosa
'Canyon Ferry'

Pinus pumila
'Dwarf Blue'

Pinus resinosa
'Globosa'
'Nobska'

Pinus strobus
'Bennett's Contorted'
'Blue Shag'
'Brevifolia'
'Contorta Nana'
'Densa'
'Dove's Dwarf'
'Green Shadow'
'Hillside Gem'
'Horsford'
'Minima'
'Minuta'
'Nana'
'Northway Broom'
'Ontario'
'Uconn'
'Verkade's Broom'

Pinus sylvestris
'Albynn's'
'Argentea Compacta'
'Beauvronensis'
'Bennett Compact'
'Bergman'
'Compact'
'Compressa'
'Doone Valley'
'Genevensis'
'Glauca Compacta'
'Glauca Globosa'
'Glauca Nana'
'Globosa'
'Globosa Viridis'
'Gold Medal'
'Grand Rapids'
'Hibernia'
'Iceni'
'Lodgehill'

'Nana Compacta'
'Nisbet's Gem'
'Pygmaea'
'Repens'
'Saxatilis'
'Sherwood'
'Tabuliformis'
'Viridis Compacta'
'Watereri'
'Watereriana'
'Windsor'

Pinus thunbergiana

'Ban-sho-ho'
'Corticosa'
'Giradi Nana'
'Nishiki'
'Thunderhead'
'Yatsubusa'

Pinus virginiana

'Nashawena'

Pinus wallichiana

'Nana'
'Umbraculifera'

Platycladus orientalis

'Athrotaxoides'
'Blijdenstein'
'Chinensis'
'Decussata'
'Dwarf Greenspike'
'Excelsa'
'Filiformis Elegans'
'Filiformis Erecta'
'Filiformis Nana'
'Fruitlandii'
'Globosa'
'Juniperoides'
'Macrocarpa'
'Monstrosa'
'Nana'
'Nana Compacta'
'Pygmaea'
'Rosedale'
'Rosedalis'
'Sanderi'

'Sieboldii'
'Tetragona'
'Triangularis'

Pseudolarix kaempferi

'Nana'

Pseudotsuga menziesii

'Densa'
'Fletcheri'
'Glauca Nana'
'Globosa'
'Hillside Pride'
'Holmstrup'
'Mucronata Compacta'
'Nidiformis'
'Parkland Dwarf'
'Pumila'
'Tempelhof Compact'
'Young's Broom'

Sciadopitys verticillata

'Knirps'

Sequoia sempervirens

'Nana Pendula'

Sequoiadendron gigantea

'Pygmaea'

Taxus baccata

'Amersfoort'
'Compacta'
'Corona'
'Decora'
'Epacrioides'
'Ericoides'
'Fastigiata Nana'
'Knirps'
'Nana'
'Nutans'
'Page'
'Paulina'
'President'
'Procumbens'
'Prostrata'
'Pumila'

'Pygmaea'
'Repandens'
'Repens'

Taxus brevifolia

'Nana'

Taxus canadensis

'Dwarf Hedge'
'Fastigiata'
'Pyramidalis'
'Stricta'

Taxus cuspidata

'Bobbink'
'Midget'
'Minima'

Taxus ×hunneweliana

'Richard Horsey'

Taxis ×media

'Fairview'
'Flushing'
'Heasleyi'
'Stovekeni'

Thuja occidentalis

'Batemannii'
'Bodmeri'
'Boothii'
'Caespitosa'
'Compacta'
'Cristata'
'Danica'
'Dirigo Dwarf'
'Dumosa'
'Ericoides'
'Filiformis'
'Froebellii'
'Globosa'
'Green Midget'
'Hetz Midget'
'Hoseri'
'Hoveyi'
'Intermedia'
'Leptoclada'

'Little Champion'
'Little Gem'
'Lombarts Dwarf'
'Nana'
'Ohlendorfii'
'Pygmaea'
'Reedii'
'Rheindiana'
'Sherwood Moss'
'Sphaerica'
'Tiny Tim'
'Umbraculifera'
'Van der Bom'
'Wareana Globosa'
'Watereri'
'Woodwardii'

Thuja plicata
  'Cuprea'
  'Hillieri'
  'Pumila'

Thujopsis dolobrata
  'Nana'

Torreya nucifera
  'Prostrata'

Tsuga canadensis
  'Abbott's Pygmy'
  'Abbott Weeping'
  'Andrews'
  'Bacon Cristate'
  'Beehive'
  'Bennett's Minima'

'Bergman's Cascade'
'Bergman's Gem'
'Bonnie Bergman'
'Brookline'
'Cinnamomea'
'Cloud Prune'
'Compacta'
'Curly'
'Curtis Ideal'
'Curtis Spreader'
'Doran'
'Dr. Hornbeck'
'Droop Tip'
'Elm City'
'Essex'
'Geneva'
'Gentsch Dwarf'
'Gentsch Dwarf
  Globe'
'Gentsch Globe'
'Globosa'
'Gracilis'
'Green Cascade'
'Greenspray'
'Hahn'
'Hancock'
'Helene Bergman'
'Heli'
'Henry Hohman'
'Hornbeck'
'Horsford'
'Horsford Dwarf'
'Hussii'
'Jeddeloh'
'Jervis'

'Kathryn Verkade'
'Kingston Hollow'
'Kingsville Spreader'
'Lewis'
'Milfordensis'
'Minima'
'Minuta'
'Nana'
'Nearing'
'Oldenburg'
'Palomino'
'Popeleski'
'Pumila'
'Pygmaea'
'Rockport'
'Ruggs Washington
  Dwarf'
'Starker'
'Stewart's Dwarf'
'Stewart's Dwarf
  Globe'
'Stewart's Gem'
'Verkade Petite'
'Verkade's Recurved'
'Von Helms'
'Von Helms Dwarf'
'Warner's Globe'

Tsuga diversifolia
  'Nana'

Tsuga heterophylla
  'Dumosa'
  'Epstein Dwarf'

Growth Rate Slow

Abies alba
  'Nana'

Abies balsamea
  'Nana'

Abies concolor
  'Conica'
  'Verkades
    Witchbroom'

Abies koreana
  'Starker's Dwarf'

Abies lasiocarpa
  'Mulligan's Dwarf'

Abies nordmanniana
  'Procumbens'

Abies pinsapo
  'Aurea'

Abies procera
  'Glauca'
  'Jeddeloh'

Cedrus deodara
  'Cream Puff'
  'Prostrata'

*Cephalotaxus
  harringtonia*
  'Nana'

Chamaecyparis lawsoniana
'Gold Splash'
'Grandi'
'Nestoides'

Chamaecyparis obtusa
'Albospica'
'Crippsii'
'Golden Christmas Tree'
'Golden Fairy'
'Golden Nymph'
'Golden Sprite'
'Junior'
'Lycopodiodes Aurea'
'Mariesii'
'Prostrata'
'Sanderi'
'Wells Special'

Chamaecyparis pisifera
'Argenteovariegata'
'Aurea Compacta Nana'
'Aureovariegata'
'Boulevard'
'Cream Ball'
'Cyanoviridis'
'Plumosa Cream Ball'
'Squarrosa Argentea'
'Squarrosa Argentea Compacta'
'Squarrosa Aurea'
'Squarrosa Cyano-viridis'
'Squarrosa Minima'

Chamaecyparis thyoides
'Andelyensis'
'Heatherbun'

Cryptomeria japonica
'Bandai-sugi'
'Compacta'
'Giokumo'
'Gyokruya'

'Lobbii Nana'
'Spiralis'
'Spiraliter Falcata'
'Yore-sugi'

Juniperus chinensis
'Robusta Green'

Juniperus communis
'Effusa'
'Suecica Pencil Point'

Juniperus horizontalis
'Blue Mat'
'Glauca'
'Pulchella'
'Venusta'

Juniperus sabina
'Jade'

Juniperus scopulorum
'Blue Haven'
'Blue Heaven'
'Blue Moon'
'Montana'
'Springbank'

Juniperus squamata
'Glassel'

Juniperus virginiana
'Nova'

Larix decidua
'Corley'

Larix kaempferi
'Pendula'

Picea abies
'Acutissima'
'Clanbrassiliana'
'Clanbrassiliana Elegans'
'Clanbrassiliana Plumosa'
'Clanbrassiliana Stricta'

'Globosa Nana'
'Hillside Upright'
'Kingsville'
'Loreley'
'Montnomah'
'Multonomah'
'Mutabilis'
'Parviformis'
'Prostrata'
'Pyramidalis'
'Remontii'
'Sherwood Gem'
'Sherwoodii'
'Wells Green Globe'

Picea glauca
'Brevifolia'
'Densata'
'Ericoides'
'Little Globe'
'Wild Acres'

Picea mariana
'Ericoides'
'Golden'

Picea orientalis
'Bergman's Repens'
'Gowdy'
'Nana Compacta'
'Nigra Compacta'
'Repens'

Picea pungens
'Compacta'
'Glauca Compacta'
'Glauca Globosa'

Pinus cembra
'Blue Mound'
'Chamolet'
'Chlorocarpa'
'Compacta'
'Compacta Glauca'
'Nana'
'Pygmaea'

Pinus flexilis
'Bergman Dwarf'
'Compacta'

'Glenmore'
'Glenmore Dwarf'
'Temple'
'Tiny Temple'

Pinus koraiensis
'Dwarf'

Pinus mugo
'Emerald Tower'
'Frisia'
'Ophir'

Pinus nigra
'Hornibrookiana'
'Pygmaea'

Pinus parviflora
'Aizu'
'Fukuzumi'
'Gimborn's Pyramid'
'Gi-on'

Pinus pumila
'Glauca'
'Hillside'

Pinus resinosa
'Quinobeguin'

Pinus rigida
'Sherman Eddy'

Pinus strobus
'Anna Feile'
'Bloomer's Dark
  Globe'
'Bloomer's Globe'
'Elf'
'Globosa'
'Greg's Form'
'Horsham'
'Julian's Dwarf'
'Pygmaea'
'Redfield Seedling'
'Umbraculifera'

Pinus sylvestris
'Albynn's Prostrate'
'Aurea'
'Aurea Nana'
'Nana'
'Riverside Gem'
'Spaan's Slow
  Column'
'Twiggy'

Pinus thunbergiana
'Compacta'

Pinus virginiana
'Pocono'

Platycladus orientalis
'Aurea Conspicua'
'Aurea Nana'
'Berckmans'
'Berckmans Golden'
'Conspicua'
'Gracillima'
'Westmont'

Pseudotsuga menziesii
'Little Jon'
'Nana'
'Pyramidata'
'Slavinii'

Taxus baccata
'Adpressa Variegata'
'Argentea Minor'
'Dwarf White'
'Repens Aurea'

Taxus cuspidata
'Compacta'
'Densa'
'Intermedia'
'Nana'
'Vermueleni'
'Visseri'

Taxus ×media
'Compacta'

'Flemer'
'Nana-grand'
'Newport'
'Pilaris'
'Pyramidalis'
'Sentinalis'
'Thayeri'
'Vermuelen'

Thuja occidentalis
'Cloth of Gold'
'Golden'
'Little Giant'
'Rosenthalii'
'Techny'
'Wareana'

Thuja plicata
'Gracilis'

Tsuga canadensis
'Abbott's Dwarf'
'Armistice'
'Baldwin Dwarf
  Pyramid'
'Barrie Bergman'
'Brandley'
'Broughton'
'Calvert'
'Coffin'
'Creamy'
'Dawsoniana'
'Doc's Choice'
'Fremdii'
'Gentsch White Tip'
'Greenwood Lake'
'Horsford Contorted'
'John Swartley'
'Laurie'
'Muttontown'
'Stockman's Compact'
'Stockman's Dwarf'
'Stranger'
'Unique'
'Upper Bank'
'Valentine'
'Wheelerville'

Tsuga caroliniana
'Compacta'
'LaBar Weeping'

Tsuga diversifolia
'Manifold'
'Medford Lake'

Tsuga mertensiana
'Mount Arrowsmith'
'Quartz Mountain'

Growth Rate Species Normal

Conical

Foliage All or Partly Yellow

Abies alba
'Aurea'

Abies concolor
'Aurea'
'Wintergold'

Abies koreana
'Aurea'

Abies nordmanniana
'Aurea'
'Golden Spreader'

Abies pinsapo
'Aurea'

Abies procera
'Aurea'

Calocedrus decurrens
'Aureovariegata'
'Intricata'

Cedrus atlantica
'Aurea'
'Aurea Robusta'

Cedrus deodara
'Aurea'
'Aurea Pendula'
'Aurea Wells'
'Golden Horizon'
'Gold Rush'
'Klondike'
'Wells Golden'

Cedrus libani
'Aurea'
'Aurea Prostrata'
'Gold Tip'

*Chamaecyparis
lawsoniana*
'Alumigold'
'Ashton Gold'
'Aurea'
'Aurea Densa'
'Blue Nantais'
'Elegantissima'
'Ellwood's Gold'
'Golden King'
'Golden Showers'
'Golden Triumph'
'Golden Wonder'
'Gold Splash'
'Gracilis Aurea'
'Grayswood Gold'
'Green Pillar'
'Hillieri'
'Howarth's Gold'
'Lombartsii'
'Luteocompacta'
'Lutescens'
'Minima Aurea'
'Moerheimii'
'Monumentalis Aurea'
'New Golden'
'President Roosevelt'
'Smithii'
'Southern Gold'
'Stardust'
'Stewartii'
'Versicolor'

'Westermannii'
'Winston Churchill'
'Yellow Transparent'

Chamaecyparis obtusa
'Aurea'
'Aurea Nana'
'Crippsii'
'Fernspray Gold'
'Gracilis Aurea'
'Kojolkohiba'
'Lycopodiodes Aurea'
'Mariesii'
'Nana Aurea'
'Nana Lutea'
'Opaal'
'Tetragona Aurea'
'Verdoni'
'Youngii'

*Chamaecyparis
pisifera*
'Argenteovariegata'
'Aurea'
'Aurea Compacta'
'Aurea Nana'
'Compacta
Albovariegata'
'Compressa Aurea'
'Filifera Aurea'
'Filifera Aurea Nana'
'Filifera
Aureovariegata'
'Flavescens'
'Gold Dust'
'Golden Mop'
'Gold Spangled'

'Juniperoides Aurea'
'Lutea'
'Lutescens'
'Minima Aurea'
'Minima Variegata'
'Nana Aurea'
'Plumosa Aurea'
'Plumosa Lutescens'
'Plumosa Nana Aurea'
'Plumosa Rogersii'
'Squarrosa Aurea'
'Squarrosa Aurea
  Nana'
'Squarrosa Elegans'
'Squarrosa Lutea'
'Squarrosa Sulphurea'
'Sulphurea'
'Sungold'

*Chamaecyparis
  thyoides*
'Aurea'
'Variegata'

Cryptomeria japonica
'Aurea'
'Aureovariegata'
'Elegans Aurea'
'Ogon-sugi'
'Sekkan-sugi'
'Sekko-sugi'
'Sekkwa-sugi'
'Variegata'

×*Cupressocyparis
  leylandii*
'Castlewellan Gold'
'Robinson's Gold'

Cupressus macnabiana
'Sulphurea'

Juniperus chinensis
'Kuriwao Gold'
'Sulphur Spray'

Juniperus communis
'Aurea'
'Ellis'

'Nana Aurea'
'Pendula Aurea'

Juniperus horizontalis
'Argentea'
'Aurea'
'Sun Spot'

Juniperus procumbens
'Golden'

Juniperus sabina
'Aureovariegata'

Juniperus squamata
'Holger'

Juniperus virginiana
'Aurea'
'Elegantissima'

Picea abies
'Argenteospicata'
'Aurea'
'Aurescens'
'Elegantissima'
'Finedonensis'
'Variegata'

Picea glauca
'Aurea'
'Dent'

Picea jezoensis
'Aurea'

Picea mariana
'Golden'

Picea omorika
'Aurea'

Picea orientalis
'Aurea'
'Aurea Compacta'
'Aureospicata'
'Compacta Aurea'
'Early Gold'
'Skylands'

Picea pungens
'Aurea'
'Lutea'

Picea sitchensis
'Aurea'

Pinus banksiana
'Watt's Golden'

Pinus cembra
'Aurea'
'Aureovariegata'

Pinus contorta
'Frisian Gold'
'Goldchen'

Pinus densiflora
'Aurea'
'Oculus-draconis'

Pinus koraiensis
'Dragon Eye'
'Oculus-draconis'

Pinus leucodermis
'Aureospicata'

Pinus mugo
'Aurea'
'Kokarde'
'Winter Gold'

Pinus nigra
'Aureovariegata'

Pinus parviflora
'Kouraku'
'Ogon Janome'
'Shi-on'

Pinus resinosa
'Aurea'

Pinus rigida
'Aurea'

Pinus strobus
  'Aurea'
  'Bennett's Dragon
    Eye'
  'Bergman's
    Variegated'
  'Hillside Winter Gold'
  'Winter Gold'

Pinus sylvestris
  'Aurea'
  'Aurea Nana'
  'Aureopicta'
  'Gold Coin'
  'Greg's Variegated'
  'Inverleith'
  'Variegata'

Pinus thunbergiana
  'Aurea'
  'Benijamone-
    kuromatsu'
  'Iseli Golden'
  'Janome-matsu'
  'Oculus-draconis'
  'Ogon-kuromatsu'

Pinus virginiana
  'Wate's Golden'

Pinus wallichiana
  'Zebrina'

Platycladus orientalis
  'Hillieri'
  'Mayhewiana'
  'Pyramidalis Aurea'
  'Tatarica'

Pseudotsuga menziesii
  'Hillside Gold'

Sequoia sempervirens
  'Variegata'

*Sequoiadendron
  gigantea*
  'Aurea'

Taxus baccata
  'Adpressa Aurea'
  'Adpressa Variegata'
  'Aurea'
  'Aurescens'
  'Dovastonii Aurea'
  'Dovastonii
    Aureovariegata'
  'Erecta Aurea'
  'Repens Aurea'
  'Semperaurea'
  'Washingtonii'

Taxus canadensis
  'Aurea'

Taxus cuspidata
  'Aurea'
  'Aurescens'
  'Bright Gold'
  'Dwarf Bright Gold'

Thuja occidentalis
  'Aurea Americana'
  'Cloth of Gold'
  'Douglasii Aurea'
  'George Peabody'
  'George Washington'
  'Goldcrest'

'Hetz Midget
  Variegated'
'Holmstrup Yellow'
'Lutea'
'Lutea Nana'
'Rheingold Beattie'
'Riversii'
'Semperaurea'
'Sudsworthii'
'Sunkist'
'Vervaenaena'
'Waxen'

Thuja plicata
  'Aurea'
  'Aureovariegata'
  'Canadian Gold'
  'Gracilis Aurea'
  'Rogersii'
  'Rogersii Aurea'
  'Semperaurescens'
  'Sunburst'
  'Sunshine'
  'Variegata'
  'Zebrina'

Torreya nucifera
  'Aurea Variegata'

Tsuga canadensis
  'Aurea'
  'Bennett's Golden'
  'Cappy's Choice'
  'Golden Splendor'
  'Silvery Gold'

Foliage All or Partly White

Abies balsamea
  'Argentea'
  'Variegata'

Abies concolor
  'Argentea'
  'Wattezii'

Cedrus deodara
  'Albospica'
  'White Imp'

*Chamaecyparis
   lawsoniana*

'Albovariegata'
'Croftway'
'Ellwood's White'
'Erecta Alba'
'Fleckalwood'
'Minima'
'Monumentalis
   Glauca'
'Nana Albospica'
'Nana
   Argenteovariegata'
'Silver Queen'

Chamaecyparis obtusa

'Albospica'
'Argentea'
'Snowkist'
'Tonia'

*Chamaecyparis
   pisifera*

'Cream Ball'
'Filifera
   Argenteovariegata'
'Filifera Variegata'
'Mikko'
'Nana Albovariegata'
'Plumosa Albopicta'
'Silver Lode'
'White Pygmy'

Cryptomeria japonica

'Albospica'
'Albovariegata'
'Knaptonensis'

'Nana Albospica'
'Okina-sugi'

Juniperus horizontalis

'Variegated'

Juniperus sabina

'Albovariegata'

Juniperus virginiana

'Albospica'

Picea abies

'Argentea'

Picea pungens

'Elegantissima'
'Goldie'
'Walnut Glen'

Pinus flexilis

'Albovariegata'

Pinus strobus

'Alba'

Pinus sylvestris

'Barrie Bergman'

Pinus thunbergiana

'Shirago-kuromatsu'
'Tigrina'
'Tura-ku Kuromatsu'
'Variegata'

Platycladus orientalis

'Argenteovariegata'

'Pumila Argenta'
'Summer Cream'

Sequoia sempervirens

'Adpressa'
'Albospica'

*Sequoiadendron
   gigantea*

'Argentea'

Taxus baccata

'Argentea Minor'
'Dwarf White'
'Variegata'

Thuja occidentalis

'Alba'
'Elegantissima'
'Sherwood Frost'
'Wansdyke Silver'

Thujopsis dolobrata

'Variegata'

Tsuga canadensis

'Albospica'
'Bergman's Frosty'
'Dwarf Whitetip'
'Frosty'
'Gentsch Snowflake'
'LaBar White Tip'
'Pendula Argentea'
'Silver Tip'
'Watnong Star'
'White Tip'

Rounded

Foliage All or Partly Yellow

Chamaecyparis obtusa
'Gold Drop'
'Golden Christmas
   Tree'
'Golden Fairy'
'Golden Nymph'
'Golden Sprite'

'Yellowtip'

*Chamaecyparis
   pisifera*

'Aurea Compacta
   Nana'

'Gold Dust'
'Mikko'
'Nana Aureovariegata'
'Snow'
'Sulphurea Nana'
'Winter Gold'

Picea abies
  'Ohlendorfii'

Pinus banksiana
  'Manomet'

Pinus sylvestris
  'Moseri'

Platycladus orientalis
  'Aurea'

'Aurea Conspicua'
'Aurea Nana'
'Berckmans'
'Berckmans Golden'
'Conspicua'
'Semperaurescens'

Taxus baccata
  'Pumila Aurea'

Thuja occidentalis
  'Aurea'
  'Cristata Aurea'
  'Ellwangeriana Aurea'
  'Globosa Aurea'
  'Golden Globe'
  'Pumila Sudworthii'
  'Rheingold'

Thuja plicata
  'Stoneham Gold'

Foliage All or Partly White

Cedrus deodara
  'Silver Mist'
  'Snow Sprite'

Chamaecyparis obtusa
  'Nana Argentea'
  'White Tip'

*Chamaecyparis
  pisifera*
  'Compacta Variegata'
  'Nana Variegata'
  'Snow'

Thuja occidentalis
  'Meinekes Zwerg'

Tsuga canadensis
  'Albopicta'
  'Gentsch'
  'Gentsch Snowflake'
  'Gentsch Variegated'
  'Gentsch White'
  'Gentsch White Tip'
  'Snowflake'

Foliage Green (Species Normal)

Abies alba
  'Globosa'

Abies balsamea
  'Globosa'

Abies nordmanniana
  'Compacta'
  'Refracta'

Abies procera
  'Blauehexe'

Calocedrus decurrens
  'Depressa'

Cedrus deodara
  'Compacta'

*Chamaecyparis
  lawsoniana*
  'Globosa'

'Pixie'
'Rijnhof'

Chamaecyparis obtusa
  'Flabelliformis'

*Chamaecyparis
  pisifera*
  'Argentea Nana'
  'Globosa'

Cryptomeria japonica
  'Bandai-sugi'
  'Globosa'
  'Kusari-sugi'
  'Spiralis'

Juniperus horizontalis
  'Andorra Compacta'

Juniperus sabina
  'Broadmoor'
  'New Blue'

Juniperus scopulorum
  'Globe'
  'Lakewood Globe'

Juniperus virginiana
  'Globosa'
  'Kosteri'

Picea abies
  'Compacta'
  'Highlandia'
  'Pseudomaxwellii'

Pinus banksiana
  'Chippewa'
  'Wisconsin'

Pinus densiflora
  'Globosa'
  'Tanyosho'
  'Tanyosho Special'

Pinus nigra
  'Globosa'

Pinus parviflora
  'Bergmanii'

Pinus strobus
  'Compacta'

Pinus thunbergiana
  'Globosa'
  'Kuro-bandaisho'

Pinus wallichiana
  'Silverstar'

Taxus cuspidata
  'Wilsoni'

Taxus ×media
  'Dutweilleri'
  'Gem'
  'Kohli'
  'Ohio Globe'
  'Taunton'
  'Wilsonii'
  'Wymani'

Thuja occidentalis
  'Canadian Green'
  'Conica Densa'
  'Green Midget'
  'Hoveyi'
  'Hudsonica'
  'Little Gem'
  'Little Giant'
  'Meckii'

  'Pumila'
  'Wagner'

Tsuga canadensis
  'Bristol'
  'Globosa'
  'Innisfree'
  'Laurie'
  'Mansfield'
  'Matthews'
  'Pendula'
  'Rock Creek'
  'Sherwood Compact'
  'Wilton Globe'

Tsuga caroliniana
  'LaBar Bushy'

Columnar

Abies alba
  'Columnaris'
  'Fastigata'

Abies concolor
  'Pendula'
  'Fastigiata'

Abies pinsapo
  'Fastigiata'

Calocedrus decurrens
  'Columnaris'
  'Greenspire'

Cedrus atlantica
  'Argentea Fastigiata'
  'Fastigiata'
  'Glauca Fastigiata'
  'Pyramidalis'

Cedrus deodara
  'Gold Strike'

*Cephalotaxus
  harringtonia*
  'Fastigiata'

*Chamaecyparis
  lawsoniana*
  'Alumii'
  'Blom'
  'Blue Plume'
  'Broomhill Gold'
  'Ellwoodii Glauca'
  'Erecta
   Aureovariegata'
  'Erecta Glauca'
  'Erecta Viridis'
  'Fletcheri'
  'Fletcher's White'
  'Grayswood Pillar'
  'Kelleriis Gold'
  'Kestonensis'
  'Killiney Gold'
  'Lane'
  'Lutea'
  'Monumentalis Nova'
  'Pembury Blue'
  'Pottenii'
  'Pyramidalis'
  'Pyramidalis Alba'
  'Robusta'
  'Robusta Glauca'
  'Snow Flurry'

  'Viner's Gold'
  'White Spot'

*Chamaecyparis
  thyoides*
  'Hopkinton'

Cryptomeria japonica
  'Monstrosa'
  'Yoshino'

*×Cupressocyparis
  leylandii*
  'Green Spire'
  'Leighton Green'
  'Silver Dust'
  'Stapehill'

Cupressus arizonica
  'Fastigiata'
  'Fastigiata Aurea'

Juniperus chinensis
  'Fortunei'
  'Mission Spire'

Juniperus communis
  'Arnold'

'Ashfordii'
'Columnaris'
'Contraversa'
'Cracovia'
'Erecta'
'Grayii'
'Hibernica'
'Kiyonoi'
'Laxa'
'Sentinel'
'Suecica'

Juniperus sabina

'Fastigiata'

Juniperus scopulorum

'Cologreen'
'Columnar Sneed'
'Columnaris'
'Dew Drop'
'Erecta'
'Greenspire'
'Hill's Silver'
'Medora'
'Moffet Blue'
'Welchii'

Juniperus virginiana

'Boskoop Purple'
'Columnaris'
'Emerald Sentinel'
'Fastigiata'
'Glauca'
'Glauca Compacta'
'Hillii'
'Pseudocupressus'
'Pyramidiformis'
'Robusta Green'
'Skyrocket'
'Sparkling Skyrocket'

Larix decidua

'Fastigiata'

Picea abies

'Columnaris'
'Cupressina'

Picea mariana

'Fastigiata'

Picea pungens

'Columnaris'
'Iseli Fastigiate'

Pinus banksiana

'Fastigata'

Pinus cembra

'Columnaris'
'Stricta'

Pinus mugo

'Rigi'

Pinus nigra

'Columnaris'

Pinus strobus

'Fastigiata'

Pinus sylvestris

'Fastigiata'
'Sentinel'

Platycladus orientalis

'Beverleyensis'
'Blue Cone'
'Columnaris'
'Maurieana'
'Meldensis'

Pseudotsuga menziesii

'Fastigiata'

Taxus baccata

'Columnaris'
'Fastigiata'
'Fastigiata Aurea'
'Fastigiata
Aureomarginata'
'Fastigiata
Aureovariegata'
'Hibernica'
'Melfard'
'Stricta'

Taxus brevifolia

'Erecta'

Taxus cuspidata

'Adams'
'Columnaris'
'Erecta'
'Pyramidalis'
'Stovekenii'

Taxus ×media

'Anthony Wayne'
'Columnaris'
'Costich'
'Erecta'
'Green Candle'
'Hicksii'
'Parade'
'Robusta'
'Totem'
'Vermuelen'
'Viridis'
'Wellesleyana'

Thuja occidentalis

'Ada'
'Brandon'
'Columbia'
'Columna'
'Douglasii
Pyramidalis'
'Malonyana'
'Sherwood Column'
'Skogholm'
'Unicorn'
'Watnong Gold'
'Wintergreen'

Thuja plicata

'Excelsa'
'Fastigiata'

Tsuga canadensis

'Columnaris'
'Fastigiata'
'Kingsville'
'Moon's Columnar'
'Schramm'
'Vermeulen's
Pyramid'

Low, Spreading over the Ground

Abies balsamea
  'Andover'

Abies concolor
  'Wattezii'

Abies fraseri
  'Prostrata'

Abies homolepis
  'Prostrata'

Abies koreana
  'Prostrata'
  'Prostrate Beauty'

Abies nordmanniana
  'Golden Spreader'
  'Prostrata'

Abies procera
  'Glauca Prostrata'
  'Prostrata'

Cedrus deodara
  'Repens'
  'Viridis Prostrata'

Cedrus libani
  'Aurea Prostrata'
  'Golden Dwarf'

*Cephalotaxus
  harringtonia*
  'Prostrata'

Chamaecyparis obtusa
  'Nana Prostrata'
  'Prostrata'
  'Repens'

*Chamaecyparis
  pisifera*
  'Pendula'
  'Sopron'

Juniperus chinensis
  'San Jose'

Juniperus communis
  'Derrynana'
  'Edgbaston'
  'Gew Graze'
  'Gimborn'
  'Gold Beach'
  'Hornibrookii'
  'Inverleith'
  'Minima'
  'Prostrata'
  'Repanda'
  'Soapstone'
  'Suecica'
  'Windsor Gem'

Juniperus conferta
  'Blue Pacific'
  'Emerald Green'

Juniperus horizontalis
  'Admirabilis'
  'Adpressa'
  'Alpina'
  'Andorra'
  'Andorra Compacta'
  'Argentea'
  'Aunt Jemina'
  'Bar Harbor'
  'Blue Acres'
  'Blue Chip'
  'Blue Mat'
  'Blue Moon'
  'Blue Rug'
  'Blue Wilton'
  'Prince of Wales'
  'Prostrata'

Juniperus procumbens
  'Golden'
  'Nana'

Juniperus scopulorum
  'Palmeri'
  'Prostrata'

Juniperus squamata
  'Blue Carpet'
  'Prostrata'

Juniperus virginiana
  'Green Spreader'
  'Prostrata'
  'Reptans'

Picea abies
  'Dumosa'
  'Inversa'
  'Procumbens'
  'Prostrata'
  'Pseudoprostrata'
  'Reflexa'
  'Repens'
  'Sargentii'

Picea alcoquiana
  'Prostrata'

Picea mariana
  'Procumbens'

Picea omorika
  'Berliners Weeping'
  'Expansa'

Picea pungens
  'Blue Spreader'
  'Glauca Procumbens'
  'Glauca Prostrata'
  'Procumbens'
  'Prostrata'
  'Prostrate Blue Mist'

Pinus banksiana
  'Neponset'
  'Schoodic'
  'Uncle Fogy'

Pinus cembra
  'Prostrata'

Pinus densiflora
  'Griffith Prostrate'
  'Pendula'

Pinus echinata
  'Clines Dwarf'

Pinus mugo
  'Corley's Mat'
  'Spaan'
  'Spaan's Pygmy'
  'Valley Cashion'

Pinus nigra
  'Prostrata'

Pinus pumila
  'Prostrata'

Pinus strobus
  'Prostrata'

Pinus sylvestris
  'Albynn'
  'Albynn's Prostrate'
  'Hillside Creeper'

Pseudotsuga menziesii
  'Prostrata'

Sequoia sempervirens
  'Cantab'
  'Prostrata'
  'Repens'

Taxus baccata
  'Cavendishii'

'Decora'
'Ericoides'
'Procumbens'
'Prostrata'
'Repandens'

Taxus cuspidata
  'Depressa'
  'Prostrata'

Torreya nucifera
  'Prostrata'

Tsuga canadensis
  'Cole'
  'Coles Prostrate'

Conifer Families and Genera and Their Distribution

Family (no. of genera) and genus	Number of species[a]	Number of cultivars[b]	Native distribution
Taxaceae (5)	23 (9)	233	
Amentotaxus	4	—	W. China
Austrotaxus	1	—	New Caledonia
Pseudotaxus	1	—	W. China
Taxus	10 (7)	231	Northern Hemisphere
Torreya	7 (2)	2	E. Asia, N. America
Cephalotaxaceae (1)	9 (2)	13	
Cephalotaxus	9 (2)	13	Honduras to E. Asia
Podocarpaceae (15)	141		
Acmopyle	3	—	New Caledonia, Fiji Islands
Dacrycarpus	10	—	Burma to New Zealand
Dacrydium	16	—	Islands of Australasia, Chile
Decussocarpus	11	—	Asia, Africa, S. America
Falcatifolium	4	—	Malaysia to New Caledonia
Halocarpus	3	—	New Zealand
Lagarostrobos	2	—	Tasmania, New Zealand
Lepidothamnus	3	—	Chile, New Zealand
Microcachrys	1	—	Tasmania
Microstrobos	2	—	Tasmania, New South Wales
Parasitaxus	1	—	New Caledonia
Phyllocladus	5	—	Borneo, Philippines, Islands of Australia
Podocarpus	75	—	Asia, Africa, S. America
Prumnopitys	4	—	S. America, New Zealand
Saxegothea	1	—	S. Chile, Patagonia
Araucariaceae (2)	31		
Agathis	13	—	Indonesia to E. Australia, New Zealand
Araucaria	18	—	S. America, Australia and islands
Taxodiaceae (10)	16 (8)	107	
Athrotaxis	3	—	Tasmania
Cryptomeria	1 (1)	70	Japan
Cunninghamia	3 (1)	2	China, Formosa
Glyptostrobus	1	—	China
Metasequoia	1 (1)	1	China
Sciadopitys	1 (1)	5	Japan
Sequoia	1 (1)	20	N. America
Sequoiadendron	1 (1)	6	N. America
Taiwania	1	—	Formosa
Taxodium	3 (2)	3	N. America
Pinaceae (10)	234 (86)	1176	
Abies	49 (23)	116	Northern Hemisphere

Family (no. of genera) and genus	Number of species[a]	Number of cultivars[b]	Native distribution
Cathaya	1	—	China
Cedrus	4 (3)	78	Himalaya, N. Africa, Mediterranean
Keteleeria	7	—	China
Larix	15 (5)	25	Northern Hemisphere
Picea	34 (15)	318	Northern Hemisphere
Pinus	100 (31)	367	Northern Hemisphere, Indonesia
Pseudolarix	1 (1)	1	N.E. China
Pseudotsuga	5 (1)	43	N. America, E. Asia
Tsuga	18 (7)	228	N. America, Himalayas to Japan
Cupressaceae (21)	128 (24)	1140	
Actinostrobus	3	—	W. Australia
Austrocedrus	1	—	S. America
Callitris	14	—	Australia, Tasmania, New Caledonia
Calocedrus (Heyderia)	3	8	N. America, S.E. Asia
Chamaecyparis	7 (5)	435	N. America, S.E. Asia
×Cupressocyparis	1 (1)	—	
Cupressus	15 (2)	7	W.N. America, E. Asia to Mediterranean
Diselma	1	—	Tasmania
Fitzroya	1	—	S. Chile, Patagonia
Fokienia	1	—	China
Juniperus	58 (10)	422	Northern Hemisphere
Libocedrus	5	—	New Zealand, New Caledonia
Microbiota	1 (1)	—	S.E. Siberia
Neocallitropsis	1	—	New Caledonia
Papuacedrus	3	—	New Guinea
Pilgerodendron	1	—	S. Chile to Patagonia
Platycladus	1 (1)	81	China, Korea
Tetraclinis	1	—	S. Spain, N. Africa
Thuja	6 (3)	185	E. Asia, N. America
Thujopsis	1 (1)	2	Japan
Widdringtonia	3	—	S.E. Africa, tropical Africa
Total 7 families, 64 genera	582 (129)	2669	

Modified from Cope, E. A. 1984. Taxonomy and conifers. Bull. Am. Conifer Soc. 1:68–71.

[a]Parenthetical numbers indicate the number of species described in this book.

[b]Number of cultivars mentioned in this book only and not the total number of cultivars.

Representative Cones and Seeds of 27 Genera of Conifers

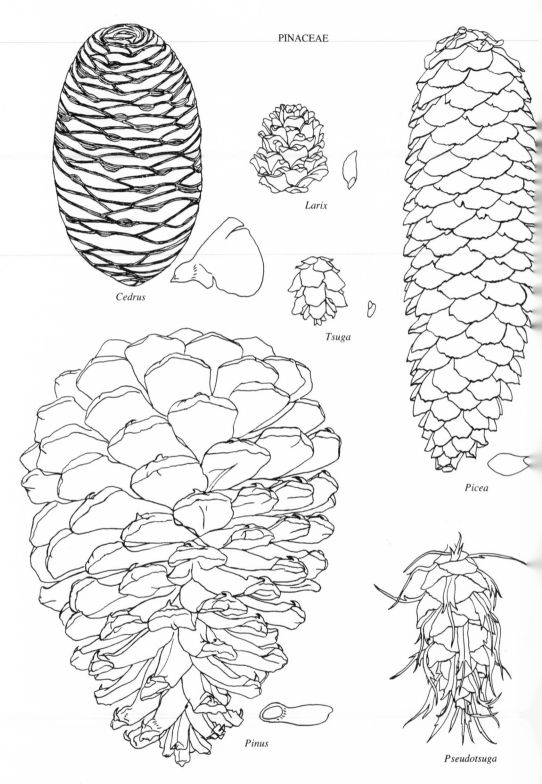

Cedrus

Larix

Tsuga

Picea

Pinus

Pseudotsuga

All drawings ×0.9.

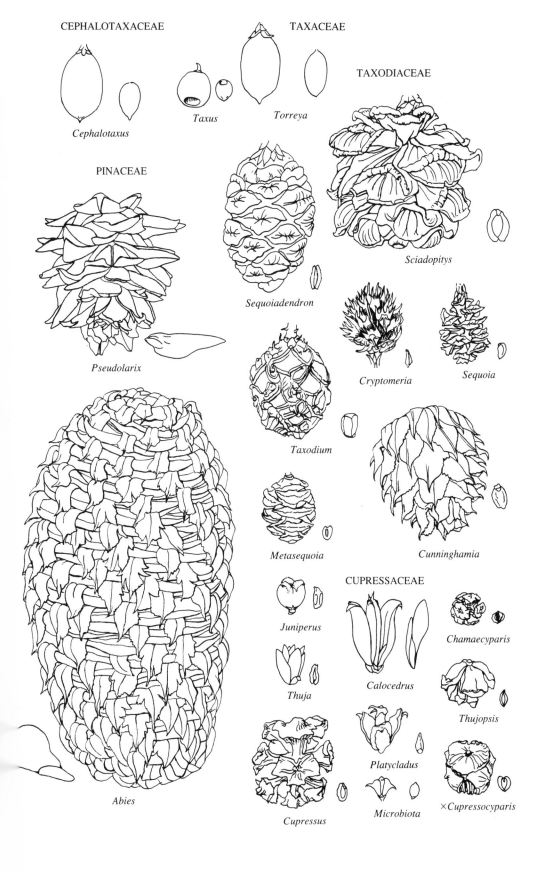

CEPHALOTAXACEAE

Cephalotaxus

TAXACEAE

Taxus

Torreya

TAXODIACEAE

Sciadopitys

PINACEAE

Sequoiadendron

Cryptomeria

Sequoia

Pseudolarix

Taxodium

Cunninghamia

Metasequoia

CUPRESSACEAE

Juniperus

Chamaecyparis

Thuja

Calocedrus

Thujopsis

Platycladus

Cupressus

Microbiota

×*Cupressocyparis*

Abies

APPENDIX 4

Conifer Twigs

PINACEAE

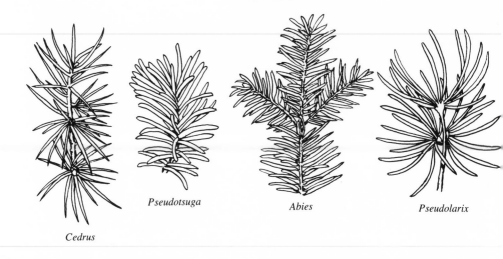

Cedrus *Pseudotsuga* *Abies* *Pseudolarix*

CUPRESSACEAE

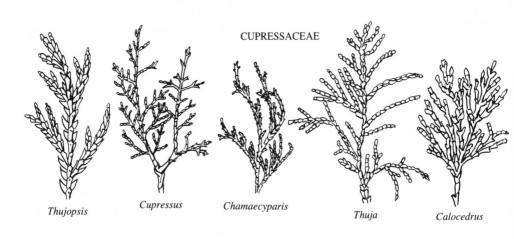

Thujopsis *Cupressus* *Chamaecyparis* *Thuja* *Calocedrus*

CEPHALOTAXACEAE

TAXODIACEAE

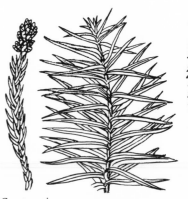

Cephalotaxus *Taxodium* *Cryptomeria* *Cunninghamia* *Metasequoia*

All drawings ×0.5.

PINACEAE

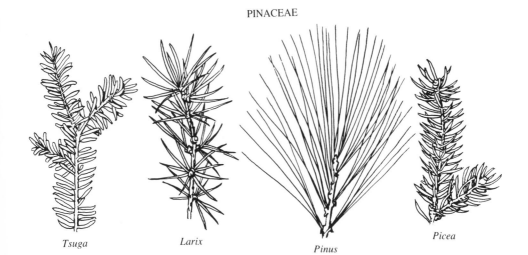

Tsuga   *Larix*   *Pinus*   *Picea*

CUPRESSACEAE

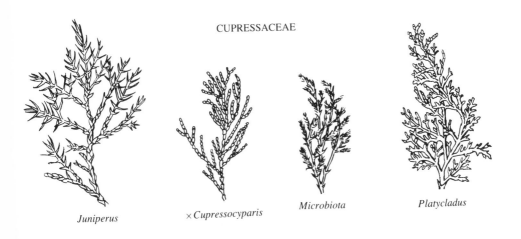

Juniperus   ×*Cupressocyparis*   *Microbiota*   *Platycladus*

TAXODIACEAE   # TAXACEAE

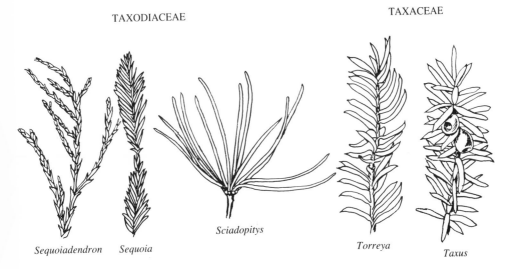

Sequoiadendron   *Sequoia*   *Sciadopitys*   *Torreya*   *Taxus*

Selected References

BAILEY, L. H. 1933. The cultivated conifers in North America. New York: Macmillan.

BAILEY HORTORIUM STAFF. 1976. Hortus III. New York: Macmillan.

BRICKELL, C. D., ed. 1980. International code of nomenclature for cultivated plants. Utrecht: Bohn, Scheltema & Hollema.

CHADWICK, L. C., AND R. A. KEEN. 1976. A study of the genus *Taxus*. Research bulletin 1086. Wooster: Ohio State Agriculture and Research Development Center.

DALLIMORE, W., AND A. B. JACKSON. 1966. A handbook of Coniferae and Ginkgoaceae. Revised by S. G. Harrison. New York: St. Martin's Press.

DEN OUDEN, P., AND B. K. BOOM. 1965. Manual of cultivated conifers. The Hague: Martinus Nijhof.

EVERITT, T. H. 1980. The New York Botanical Garden illustrated encyclopedia of horticulture. New York: Garland.

HARRISON, C. R. 1984. Ornamental conifers. Beaverton, Oreg.: Timber Press.

HILLIER, H. G. 1977. Hillier's manual of trees and shrubs. Newton Abbot: David and Charles.

HORNIBROOK, M. 1938. Dwarf and slow-growing conifers. London: Country Life and Newnes.

KRÜSSMANN, G. 1979. Die Nadelgehölze. Berlin: Paul Parey.

_____. 1983. Handbuch der Nadelgehölze. Berlin: Paul Parey.

LITTLE, E. L. 1979. Checklist of United States trees. Agriculture Handbook 541. Washington, D.C.: U.S. Department of Agriculture.

LIU, T. S. 1971. A monograph of the genus *Abies*. Taiwan: Taiwan University, Department of Forestry.

MITCHELL, A. 1974. A field guide to the trees of Britain and northern Europe. Boston: Houghton Mifflin.

REHDER, A. 1927. Manual of cultivated trees and shrubs. New York: Macmillan.

SWARTLEY, J. C. 1985. The cultivated hemlocks. Beaverton, Oreg.: Timber Press.

VALAVANIS, W. N. 1976. Japanese five-needle pine. Atlanta, Ga.: Symmes Systems.

WELCH, H. J. 1979. Manual of dwarf conifers. New York: Theophrastus.

Index

Page numbers for illustrations are shown in boldface type and are listed under the scientific name. Readers familiar only with the common name should check the pages listed for the common name to find the scientific name.